TC 31-73

Special Forces Advisor Guide

DESTRUCTION NOTICE: Destroy by any method that will prevent disclosure of contents or reconstruction of the document.

Headquarters, Department of the Army

July 2008

TRAINING CIRCULAR

No. 31-73

TC 31-73

SPECIAL FORCES ADVISOR GUIDE

2008

TC 31-73
SPECIAL FORCES ADVISOR GUIDE
2008

ISBN-13: 978-1481835558

ISBN-10: 1481835556

Proudy Printed in the
U.S.A

Training Circular
No. 31-73

Headquarters
Department of the Army
Washington, DC, 2 July 2008

Special Forces Advisor Guide

Contents

Figures

Tables

Preface

Training Circular (TC) 31-73, *Special Forces Advisor Guide*, supports Field Manual (FM) 3-05.20, *(C) Special Forces Operations (U)*—the keystone manual of the United States (U.S.) Army Special Forces (SF)—as well as FM 3-05.202, *Special Forces Foreign Internal Defense Operations.* It defines the subjective, intangible nuances of human interaction. It is designed to assist the SF Soldier in understanding and navigating the complexities of human behavior as it relates to cross-cultural communication.

PURPOSE

As with all doctrinal manuals, TC 31-73 is authoritative but not directive. It serves as a guide, but does not preclude SF units from developing their own standing operating procedures (SOPs) to meet their needs. This TC focuses on interface between SF, United States Government (USG) personnel, foreign government personnel, intergovernmental organizations (IGOs), and nongovernmental organizations (NGOs); however, the content is intended to prove useful in any human interaction.

SCOPE

The primary audiences of this TC are the commanders, staff officers, and operational personnel of Special Forces units, particularly Special Forces operational detachments A (SFODAs), Special Forces operational detachments B (SFODBs) and special operations task forces (SOTFs). Although useful to Civil Affairs (CA) and Psychological Operations (PSYOP) personnel dealing with foreign nationals and OGAs, this TC is intended specifically for SF personnel acting as advisors or liaisons. It is not designed to be a comprehensive reference manual or training guide for any specific doctrinal mission. When applied appropriately, the information and techniques are extremely useful in forming and maintaining relationships with U.S. and foreign counterparts across the spectrum of the SF core tasks.

APPLICABILITY

Commanders and trainers should use this and other related manuals in conjunction with command guidance and country- or region-specific cultural knowledge to plan and conduct successful advisor and liaison operations. This TC is applicable to the Active Army, the Army National Guard (ARNG)/Army National Guard of the United States (ARNGUS), and the United States Army Reserve (USAR) unless otherwise stated.

ADMINISTRATIVE INFORMATION

The proponent of this TC is the United States Army John F. Kennedy Special Warfare Center and School (USAJFKSWCS). Submit comments and recommended changes to Commander, USAJFKSWCS, ATTN: AOJK-DTD-SF, Fort Bragg, NC 28310-9610. This TC is designed to be UNCLASSIFIED in order to ensure the widest distribution possible to the appropriate Army special operations forces (ARSOF) and other interested USG agencies while protecting technical or operational information from automatic dissemination under the International Exchange Program or by other means. Unless this publication states otherwise, masculine nouns and pronouns do not refer exclusively to men.

Chapter 1

Special Forces as Advisors

The U.S. Army benefited greatly from foreign advisors who came to America during the Revolution to serve in the Continental Army. However, none contributed quite as significantly as Major General Baron Fredrick von Steuben. His contributions as General George Washington's Inspector General of the Army instilled discipline and professionalism into an army that previously lacked formalized training. His drill manual, taken from the Prussian army, was the backbone of the Continental Army throughout the Revolutionary War. As a benefactor of advisors such as von Steuben, the U.S. Army has since undertaken the role of advisor on numerous occasions throughout its long and illustrious history.

SF Soldiers frequently serve as advisors. The advisory role is fundamental to successfully performing the SF core tasks of unconventional warfare (UW) and foreign internal defense (FID). The demand for SF Soldiers to perform as advisors continues to grow and gain visibility.

The success enjoyed by SF advisors may be attributed directly to the cross-cultural skills gained during training and through multiple deployments and operations. The capability of SF Soldiers to perform effectively as advisors in multinational, joint, interagency, and interdependent environments represents a unique and increasingly important contribution. As such, it is likely that SF will serve as combat advisors with increasing frequency. This chapter discusses the development of SF advisors through UW, FID, counterinsurgency (COIN), and coalition warfare, and describes the peculiarities of each mission environment.

ROLES OF THE ADVISOR

1-1.   The SF military advisor has three roles involving different responsibilities and arousing differing loyalties which at times conflict. First and foremost, the advisor is a member of a U.S. military organization with a well-defined chain of command and familiar responsibilities. Within this organization, he receives and executes the orders of his superiors (which may not always be in accord with the orders his counterpart receives from his superiors). He supervises subordinate advisors. Among other duties, he must act unobtrusively (but nonetheless positively) as an inspector general—observing, evaluating, and reporting on the performance of his counterpart and the unit to which he is attached.

1-2.   Secondly, the SF advisor wears the shoulder insignia of the unit he advises, both figuratively and, quite often (as in Vietnam), literally. Living, eating, and working with the officers and men of his host unit, the SF advisor soon regards himself as one of them. The sharing of common hardships and dangers forges between him and his native counterpart potent emotional ties. The success and good name of his unit become matters of prime and personal importance to the advisor.

1-3.   Finally, the advisor is interpreter and communicator between his counterpart and his U.S. superiors and subordinates. He must introduce and explain one to the other, help resolve the myriad of problems, misunderstandings, and suspicions which arise in any human organization, particularly when people of starkly different cultures approach difficult tasks together. As has been demonstrated often, the SF advisor who has quick and easy access to an influential counterpart can sometimes be the best possible means of communicating with him.

1-4. For an advisor to be effective, he obviously must gain his counterpart's trust and confidence. This relationship, however, is only a prelude to the advisor's major objective: inspiring his counterpart to effective action. In pursuing this goal—constantly, relentlessly, and forcefully, yet patiently, persuasively, and diplomatically—the advisor must recognize conditions which can benefit or handicap his cause.

ADVISOR SCREENING AND SELECTION

1-5. Not every Soldier is well suited to perform advisory functions. Although certain individuals seem to instinctively possess the requisite skill set, others must undergo extensive interpersonal training. The selection and training process for SF is designed to optimize Soldiers' inherent abilities through a variety of training and operational experiences. Because advisors operate in very subjective environments, it is difficult to establish objective criteria by which to assess potential advisors. However, research and experience indicate that there are several personality traits that greatly enhance the SF Soldier's ability to adapt and thrive in a foreign culture. These traits include—

- Tolerance for ambiguity.
- Realistic goal and task setting.
- Open-mindedness.
- Ability to withhold judgment.
- Empathy.
- Communicativeness.
- Flexibility.
- Curiosity.
- Warmth in human relations.
- Motivation of self and others.
- Self-reliance.
- Strong sense of self.
- Tolerance for differences.
- Perceptiveness.
- Ability to accept and learn from failure.
- Sense of humor.

1-6. Of the traits listed above, no single trait is paramount. SF selection and training programs seek to develop an understanding of the contributions of each of them.

SPECIAL FORCES ADVISORS IN UNCONVENTIONAL WARFARE

1-7. Since the deployment of Jedburgh teams to occupied France during World War II (WWII), SF Soldiers have been cast into advisory roles. Conducted by, with, or through irregular forces, early UW focused on guerrilla groups operating in nonpermissive environments. SF Soldiers operating with guerrilla forces lacked institutional or positional authority. As such, these advisors had extremely limited leverage. They often were required to establish credibility by proving that their potential contribution to the surrogate force exceeded the risks associated with their presence. These advisors depended upon their persuasive personalities and cross-cultural skills to influence outcomes and ensure personal survival. The ability to influence in the absence of authority is a distinguishing quality of the successful SF advisor. As UW scenarios continue to evolve, these skills become increasingly important.

ADVISORS IN FOREIGN INTERNAL DEFENSE

1-8. In keeping with the premise that solutions to internal problems must come from internal institutions, the United States developed the concept of FID to support nations faced with internal instability. This interagency concept stresses U.S. support as part of an indigenous program designed to counter the causes of instability, such as lawlessness, subversion, or insurgency. Because of their UW- and FID-derived cross-cultural communication and advisory skills, SF have become a mainstay of U.S. FID efforts.

1-9. Throughout the Cold War, SF were called upon to implement their UW skills in actions spanning from Vietnam to El Salvador. Although SF knowledge of guerrilla warfare (GW) was important in establishing their preeminent role in these counterinsurgent operations, their UW-derived advisory skills were equally important. U.S. doctrine recognizes that long-term victory against an insurgency cannot be externally imposed; it must be accomplished through legitimate indigenous governments and forces.

1-10. Cold War COIN environments generally were more permissive than WWII UW operations, and the security of the U.S. advisor was far less dependent on the indigenous force; however, the ability to leverage counterpart cooperation remained a difficult and critical task. Although coercion may prompt surrogate forces to perform a specific action, it can never lead to institutionalization of proper conduct and legitimacy. Lasting, long-term success may only be achieved through the patient application of influence.

1-11. Operations in El Salvador provide noteworthy examples of SF advisors applying mature COIN doctrine. Employing minimal leverage and extremely restrictive rules of engagement (ROE), SF advisors played a critical role in defeating the Communist insurgency by conducting operations by, with, and through their Salvadoran military counterparts. The El Salvador model continues to be used today. In Colombia, for example, the planning assistance training team (PATT) has evolved from a single SF Advisor to a 65-man advisory assistance program that incorporates SF and conventional Army forces, as well as Air Force and Marine Corps personnel.

ADVISORS IN COALITION WARFARE

1-12. The post–Cold War era generated a corresponding reliance on coalitions—many being formed very rapidly. Coalitions present unique challenges in generating unity of effort and may be accompanied by distrust among members. More specifically, member nations are often reluctant to surrender autonomy to a single chain of command. In order to facilitate operational integration and enhanced control of these coalitions, Special Forces liaison elements (SFLEs) have been employed with non-U.S. military organizations. These liaison elements possess the skills to advise and influence coalition units, thereby enhancing the control exercised by the coalition command structure. SFLEs also facilitate operational integration of the total force. Although technical (for example, communications) and tactical (for example, coordination of air support) capabilities are instrumental to their success, it is the cross-cultural and advisory capabilities of SF that provide the greatest influence.

SPECIAL FORCES AND INTERDEPENDENT OPERATIONS

1-13. Military operations are rapidly evolving beyond joint, interagency, or multinational characterizations and toward interdependency. Interdependent operations require near-seamless integration of agencies representing all the elements of U.S. national power—diplomatic, informational, military, and economic. This same degree of integration must be extended to include such external organizations as NGOs, IGOs, and those agencies representing the various elements of power in partner states.

1-14. SF advisors commonly operate in complex joint, multinational, and interagency environments. As such, these Soldiers generally are more experienced in integrating the diverse participants in interdependent operations than their conventional counterparts. Indeed, the negotiation and cross-cultural skills that enhance their performance as advisors are also well suited to the complex interrelationships that characterize such operations. SF is uniquely suited to "operate in the seams" between these interdependent actors. Although a portion of this suitability can be attributed to organization and equipment, the advisory and liaison skills resident in SF Soldiers account for most of their effectiveness.

ORGANIZATIONAL RELATIONSHIPS

1-15. Although organizational relationships may be defined, they can often be misleading and must be clarified. The actual interrelationships between and within organizations seldom follow a line-and-block diagram. Instead, they are heavily influenced by circumstances, personalities, perceptions, and resources. All relationships and lines of authority are subject to negotiation. SF Soldiers must understand that the defining of roles, functions, and responsibilities is a continuous process, and that each statement or action sets a precedent for future interrelationships.

ORGANIZATIONAL CULTURE

1-16. SF Soldiers should approach external organizations as they would foreign counterparts—as unique institutions with distinctive cultures. Organizational differences, biases, and approaches must be researched and—whenever possible—accommodated. Other government agencies (OGAs), IGOs, and NGOs often have aims and methods that greatly differ from SF units. Within the Department of Defense (DOD), each Service has unique perspectives that lead to significant differences in objectives and approaches to various problems. Profound differences in perspective may even be found between the different branches of the U.S. Army. These characteristics, if not recognized and accommodated, may create distrust, disharmony, and disruption during critical operations. For example, the internal organizational process of the Department of State (DOS) is often described as a culture of negotiation. In this environment, the current situation is deemed a starting point, and the details of the end state are left to be determined through the negotiating process. This thought model runs counter to the traditional U.S. Army approach of first defining the end state and then backward planning, resourcing, and sequencing actions over time to achieve the predetermined objective.

1-17. The SF Soldier must become hypersensitive to the most minor organizational disparities. The apparent similarities—the common language, sociological backgrounds, and citizenship—can be disarming and lead the SF Soldier to overlook very real differences. If ignored, these differences quickly can become impediments to synchronization.

COMPARTMENTALIZATION

1-18. SF Soldiers must—as described in the special operations forces (SOF) imperatives—*balance security and synchronization.* Overcompartmentalizing must be avoided. Often distrust can be prevented simply by not being overly secretive. Although SF Soldiers must conduct a careful, continual risk assessment to avoid compromise, the sharing of critical information is necessary for harmonious operations. Perhaps the chief complaint of nonmilitary agencies is that their military counterparts continuously demand information without sharing data in return.

Chapter 2

Culture and Communication

Culture plays a crucial role in premission planning and the development of area assessments. During the course of his career, the SF Soldier travels to numerous countries and experiences a wide variety of cultures. Prior knowledge of cultural differences aids in building effective relationships and prevents embarrassment, loss of rapport, and compromise of the mission. This chapter discusses the various aspects of culture and references the readers understanding of contemporary U.S. culture to provide a framework for analysis and comparison of foreign cultures for operational purposes.

Note: Statements made in this chapter regarding particular groups of people are broad generalizations. Although these generalizations are believed to be largely valid, the intent is not to create stereotypes of different groups. It is always prudent to acknowledge the uniqueness of individuals and subgroups when analyzing any foreign culture.

THE IMPORTANCE OF CULTURE

2-1. SF teams and advisors often occupy very sensitive positions. The commander acts as the direct representative of the USG. As such, the commander and his team must be diplomats as well as military advisors. Appendix A provides checklists useful to the SF advisor. Appendix B outlines important cultural considerations.

2-2. In considering cultural awareness, SF Soldiers must observe the first SOF imperative: *Understand the operational environment.* Special operations (SO) missions require deployment outside the continental United States (OCONUS) to operate with indigenous populations whose language and culture may be very foreign to most Soldiers. SF Soldiers derive their effectiveness in large part from their ability to understand and work with foreign counterparts. This capability depends on cultural awareness.

THE DEFINITION OF CULTURE

2-3. Merriam-Webster's Collegiate Dictionary (Eleventh Edition) defines culture as: "The integrated pattern of human knowledge, belief, and behavior that depends upon the capacity for learning and transmitting knowledge, belief, and behavior to succeeding generations; the customary beliefs, social forms, and material traits of a racial, religious, or social group."

2-4. In brief, culture is the set of opinions, beliefs, values, customs, and mores that defines the identity of a society. It includes social behavior, language, and religion. Culture is a learned behavior. For example, food is a basic need that is not based on culture; however, how a person cooks and what, when, and how they eat are all products of their cultural environment.

2-5. Culture is adaptive; the customs that a group develops are based largely on a particular environment. Culture is integrated into every facet of a society. It is neither a random jumble of quaint customs nor a laundry list of dos and don'ts that a traveler should know. The physical environment and mass media are two of the strongest driving forces in changing a culture. Opinions may change fairly quickly, whereas beliefs are slow to change. Basic values are the slowest to change.

THE CONCEPT OF CULTURE

2-6.   The fact that people *learn* culture is probably culture's most essential and defining feature. Many facets of human life are transmitted genetically. An infant's desire for food, for example, is triggered by physiological characteristics determined within the human genetic code. An adult's specific desire for milk and cereal in the morning, on the other hand, is a learned (that is, cultural) response to morning hunger. The notion that milk and cereal is a breakfast food and largely inappropriate for other meals is a product of culture. As a body of learned behaviors common to a given society, human culture acts like a template, shaping behavior and consciousness within a society from one generation to the next. All learned behavior is in some degree determined by culture. The concept of a shaping template and body of learned behaviors can be broken down into the following categories:

- *Systems of meaning*, of which language is primary.
- *Ways of organizing society*, from kinship groups to states and multinational corporations.
- *Distinctive techniques practiced by a group*, and the characteristic products of these techniques.

2-7.   Several important principles follow from this concept of culture. Systems of meaning (or meaning systems) consist of negotiated agreements between members of a human society regarding the relationships between words, behaviors, or other symbols and their corresponding significance or meanings. Because meaning systems involve relationships that are not essential and universal (for example, the word "door" has no essential connection to the physical object; people simply agree to call that object a "door" in English), different human societies will inevitably arrive at different relationships and meanings.

2-8.   Just as the process of learning is an essential characteristic of culture, the process of teaching is a crucial component. Because the relationship between what is taught and what is learned is not absolute— that is, some of what is taught is lost and new innovations are introduced—*culture is in a constant state of change*. Throughout most of human history cultural changes occur quite slowly, taking place over a period of many years in a series of slow-moving accommodations to new circumstances. However, in recent times a number of rapid cultural changes have been introduced. The attempt to assimilate these rapid changes often creates tension (or even violence).

CULTURE AND THE PHYSICAL ENVIRONMENT

2-9.   The environment is the foundation of culture. The Bedouins of the Arabian deserts provide a classic example of how environment shapes culture. In a constant search for water, the Bedouins evolved culturally as nomadic herders of camels and goats. The nomadic nature of their society conditions much of their belief system, their social structure, and even their diet. Because a nomadic people cannot sustain positional warfare, their concept of war is a series of raids and skirmishes.

2-10. Learned culture serves as a lens that filters all incoming information. More importantly, it helps individuals decide which information is important and what it means. The system of meanings that constitutes a culture can dramatically influence the way people perceive and make sense of the physical world around them. Information passes through this lens of culture and is filtered, or interpreted, into a recognizable pattern. Human perspective regarding the surrounding environment is a good example of this. The same physical object or environmental element can have widely divergent meanings when perceived through different cultural filters, as described in the following examples.

2-11. In the Judeo-Christian cultural tradition, rain served as a tool of God's wrath. Although certain Biblical narratives describe rain as a blessing, the Old Testament story of Noah and the flood is familiar to nearly everyone in the Western World. Consequently, it has been very influential in establishing the place of rain and storm imagery in the cultural meaning system. One result of this system of meaning is that images like thunder, lightning, wind, and downpours tend to symbolize ominous things like anger, danger, and hardship in cultures that descend from the Judeo-Christian tradition. Examples of this effect can be found throughout Western legends and literature.

2-12. Unlike the Judeo-Christian culture, the Anasazi people believed rain to be sacred. The Anasazi lived in the Four Corners region of the U.S. Southwest (Arizona, Colorado, Utah, and New Mexico), an arid high-desert environment. Reliant upon the winter snow in the mountains to feed streams and springs throughout the year, and reliant on spring rain showers to supplement their strenuous irrigation efforts, the Anasazi culture came to view rain as a gift from the Rain God. As a result of this system of meaning—in which rainfall symbolized the benevolence and generosity of the natural world—Anasazi tradition tends to represent rain in a positive way.

2-13. Regardless of the region, these systems of meaning operate as a filter, giving significance and meaning to the group members' perceptions of the physical world around them.

CULTURAL REGIONS

2-14. The concept of cultural regions is widely accepted; however, there remains some disagreement about the number of regions and exactly what each region comprises. Countries and peoples within each region tend to have similar governments, economic systems, languages, religions, social organizations, populations, and resources. The boundaries of these regions are neither permanent nor rigid. For example, although the countries of Eastern Europe are clearly part of the European cultural region, their lengthy domination by the Union of Soviet Socialist Republics (USSR) left them with certain characteristics more typical of Russia and the Independent Republics. For purposes of this TC, the world is divided into the following seven cultural regions:

- North America and Europe (including Australia and New Zealand).
- Southwest Asia and North Africa.
- South and Central America (including Mexico).
- Sub-Saharan Africa.
- Pacific Rim (excluding the Americas).
- Russia and the Independent Republics.
- Oceania (the Pacific islands).

SUBREGIONS

2-15. Each of the major regions listed above may be further divided into subregions. There may be two large, very distinct culture groups within a single region or even a single country. South Africa, for example, has large populations of both European and sub-Saharan African cultures. Likewise, Australia has large populations of both European and Aboriginal cultures. Other regions may have numerous cultural subregions. For example, South America may be divided into cultural subregions classified as tropical-plantation, European-commercial, Amerind-subsistence, and Mestizo-farming.

DISCONTINUITY

2-16. Regions may be geographically discontiguous—Australia and New Zealand are considered to be in the European cultural region even though they are located on the opposite side of the globe. Additionally, some geographically connected members attempt to separate themselves from their cultural region. For example, Brazil is part of the South and Central American culture region; however, Brazilians do not perceive themselves as Hispanics. The reason for this is that Brazil was a former colony of Portugal and not Spain. Therefore, Brazilians are from a Lusitanian—and not a Hispanic—tradition. Although this may appear a minor point to certain outsiders, it is considered a very important distinction to Brazilians. As such, Brazilians may take offense at being referred to as Hispanic or being addressed in Spanish.

SUBCULTURES

2-17. Additionally, all cultural regions have marginalized subcultures that are typically quite different and distinct from the majority culture. The Montagnards in Southeast Asia and Native Americans in the United States are two such subcultures. Often ignored, or actively persecuted by the local government, these types of subgroups may be particularly important to the SF advisor. Historically, SF personnel often found themselves working with members of such marginalized groups.

SIGNIFICANCE

2-18. Recognition of culture is important because humans are social beings and have, throughout history, come together in groups. These groups have developed unique characteristics. Experience teaches that the more that people understand about other people and places, the more they can enrich their own culture and the less likely they are to blunder into conflict. As the world becomes increasingly more accessible, SO are becoming more dependent on the ability of the special advisor to demonstrate an understanding of the rest of the world.

KLUCKHOHN MODELS

2-19. Figures 2-1 through 2-7, pages 2-4 through 2-7, are simplifications of the Kluckhohn Value Orientations Method developed by Florence Kluckhohn and Fred Strodtbeck of Harvard University, two pioneers in the field of cultural anthropology. They are generally referred to simply as Kluckhohn Models. As presented here, they represent sweeping generalizations about very large regions. They are deliberate simplifications, intended only to capture some of the basic cultural differences and similarities among cultural regions.

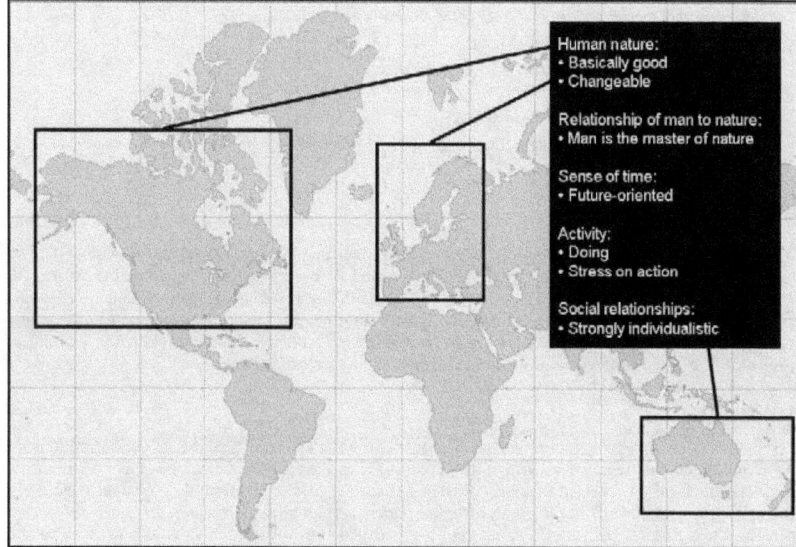

Figure 2-1. Culture analysis—North America and Europe
(including Australia and New Zealand)

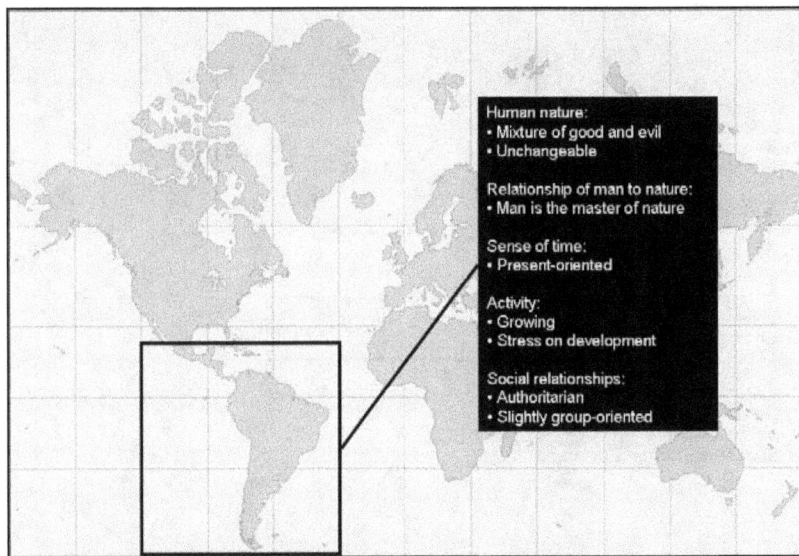

Figure 2-2. Culture analysis—South America and Central America (to include Mexico)

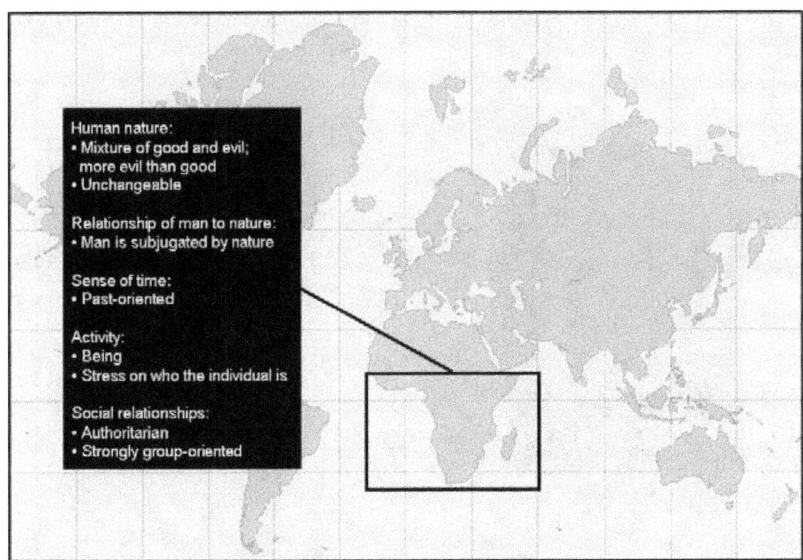

Figure 2-3. Culture analysis—Sub-Saharan Africa

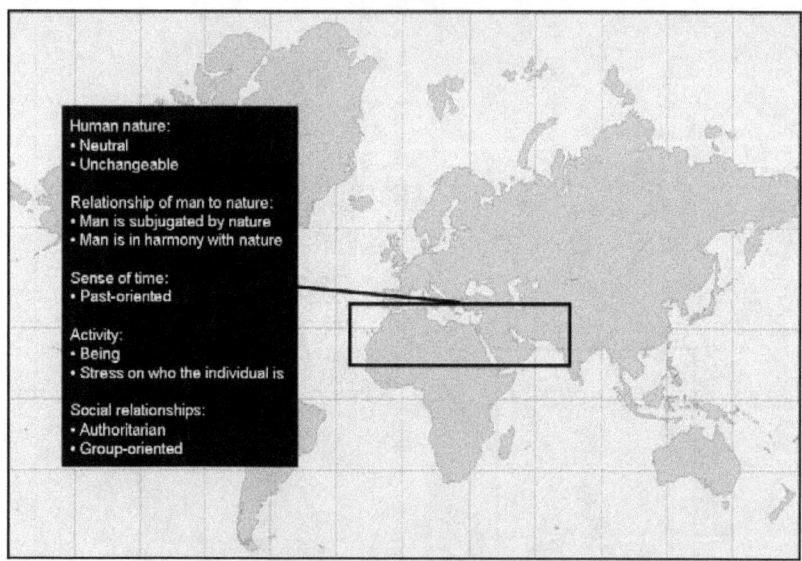

Figure 2-4. Culture analysis—North Africa and Southwest Asia

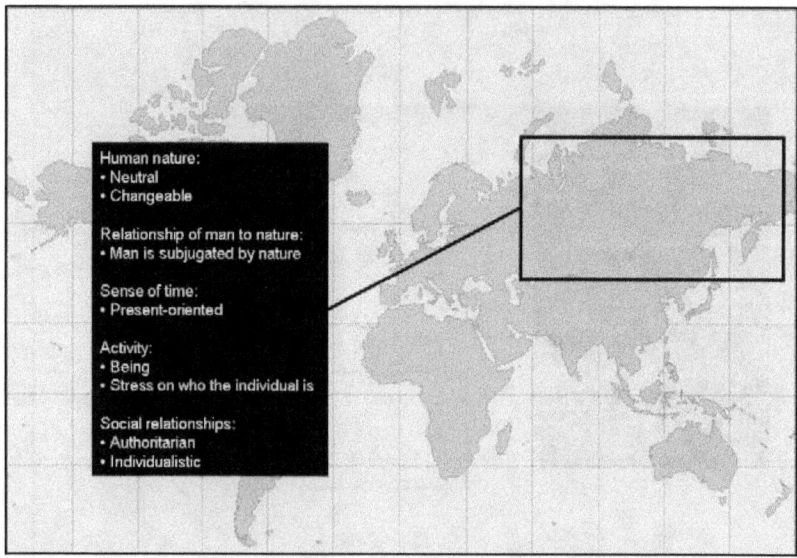

Figure 2-5. Culture analysis—Russia and the Independent Republics

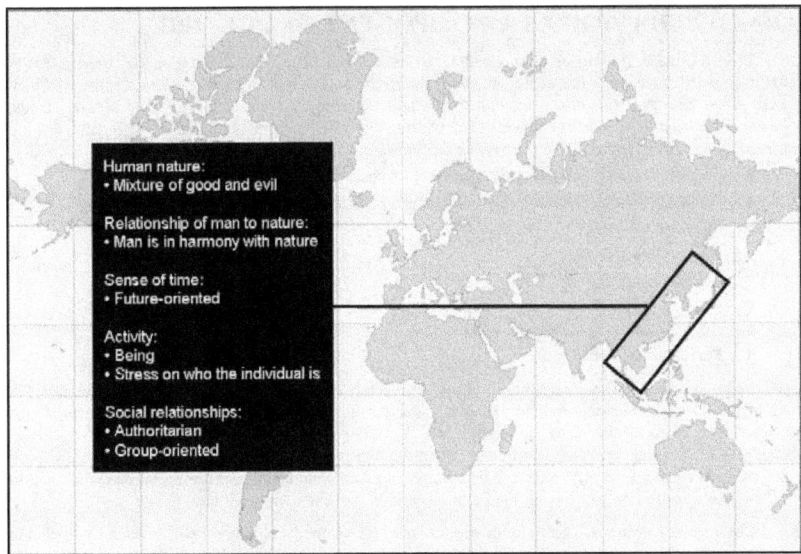

Figure 2-6. Culture analysis—Pacific Rim

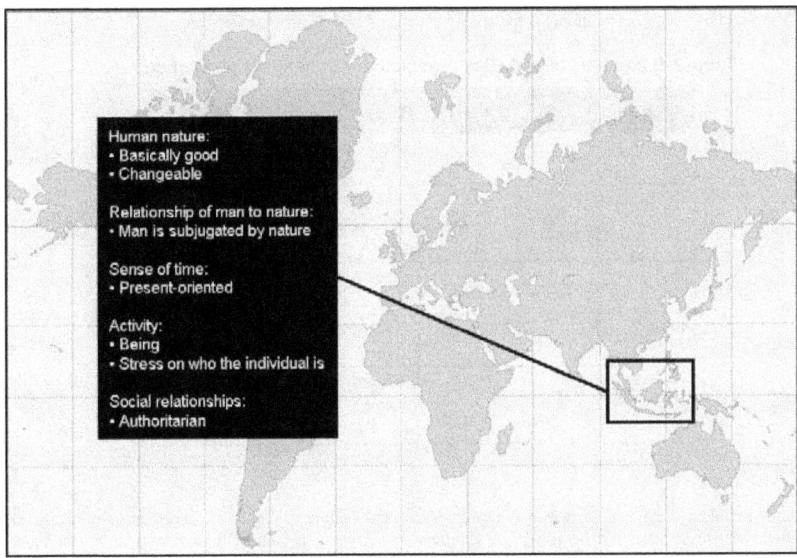

Figure 2-7. Culture analysis—Oceania

LEARNED BEHAVIOR AS A COMPONENT OF CULTURE

2-20. The baseline definition of culture indicates that learned behaviors are an essential component of culture. Learned behavior in this sense can mean almost anything—the way a person dresses, the way they speak, or the food they eat. Whenever individuals brush their teeth, cross their legs, send birthday cards, kiss someone, listen to music, or choose a form of recreation, they are practicing learned behaviors that are a part of their culture. These learned behaviors encompass all aspects of life and include attitudes toward—

- Specific people (such as family members, lovers or spouses, teachers, and friends).
- Authority figures (such as military superiors or police officers).
- Small social groups.
- Large social groups.
- Eating and food preparation.
- Work or production.
- Home building, maintenance, and cleaning.
- Recreation and relaxation.

2-21. Behaviors do not always define a culture. Frequently, there are more telling cultural signs in the meaning attached to a behavior than in the actual behavior. For example, one individual may use marijuana in Amsterdam while another does the same in New York City. Their behavior may be identical; however, the significance of that behavior is different in Amsterdam (where marijuana is legal) than in New York City (where it is not). Similarly, although many modern urban Americans hunt, the significance of hunting for them is very different than it is for the Eskimo cultures above the Arctic Circle.

2-22. The meaning systems that provide the context for learned behaviors overlap broadly with the meaning systems that constitute a society's values and beliefs. This can be illustrated by the meanings assigned to various forms of recreation in American society. Recreation is not only a set of pleasant pastimes; it has meanings and associations attached that go well beyond the simple activity itself. Table 2-1 illustrates how certain recreational activities are associated with typical participants.

Table 2-1. Example of activities associated with different parts of society

Activity	Participants
Polo	Upper class (including royalty) Male
Skateboarding	Blue or white collar Teen or preteen Male
Golf	White collar Male and female
Bowling	Blue collar Male and female
Gymnastics	White collar Young Female

BELIEFS

2-23. Beliefs can be said to take two forms: values and opinions. Values are broad moral conclusions about the way life should be lived. They reflect strong personal beliefs drawn from faith, childhood environments, and life experiences. They lend shape and order to people's lives, helping them to decide if they settled on the right actions, words, behaviors, friends, mates, leaders, religions, and careers. Values are not easily changed. Opinions are immediate conclusions about some aspect of the environment. They are usually quite specific and are susceptible to change.

2-24. Values and opinions are important because they guide the way that individuals deal with the world around them. As deeply held beliefs about rightness, values shape how individuals think things should be. Values may never be articulated—in fact, they may not even be consciously held. At times, the beliefs of an individual may even contradict one another.

2-25. Many cultures believe that their internal values are universal—that everybody knows right from wrong as *they* perceive it. Figure 2-8 depicts 20 values or qualities that are admired by most Americans. Those who do not accept these values often are viewed as deluded, perverse, or evil. The SF advisor must strive to avoid this mindset in order to be successful. Advisors often must work with counterparts with different or opposing values. Any time spent trying to change a counterpart's basic values is time wasted; furthermore, such efforts are likely to create lasting resentment.

Preparation for the Future	Originality
Cleanliness	Physical Power
Wariness	Aggressiveness
Frugality	Persistence
Timely Action	Responsibility
Practicality	Sportsmanship
Punctuality	Physical Appearance
Diligence	Hard Work
Thrift	Privacy
Initiative	Presentation of Self

Figure 2-8. Twenty basic American values or qualities

2-26. Advisors should have at least some understanding of a counterpart's value system. Values are usually nonnegotiable; they are basic rules about what is right and wrong. In dealing with a counterpart who has conflicting values, it is typically best to persuade the individual that the advocated action or behavior is not really in conflict with their values.

2-27. The systems of values and opinions that are characteristic of a culture overlap significantly with the other components of culture. These systems determine the cultural significance of even a recreational pursuit like bowling (in the learned behavior component), and how religious belief systems can intertwine with systems of social organization like political government.

2-28. Opinion/value systems overlap so much with these other components of cultural systems largely because opinions and values play such a pervasive role in culture. The baseline definition of culture suggests that culture consists essentially of learned behaviors and the template-effect by which a growing, changing culture is passed on from generation to generation. Because beliefs affect virtually every learned behavior, the cultural template consists to a significant degree of belief systems. Thus, belief systems are a central component of the larger cultural systems in which they exist.

2-29. Belief systems involve stories and myths. The interpretation of these stories and myths provides people with insight into how they should feel, think, and behave. The elaborate polytheistic mythologies of the ancient Greek and Roman civilizations are good examples of belief systems that influenced the daily lives of the societies' members. The most prominent systems of beliefs tend to be those associated with formal religions; however, any system of belief in which the interpretation of stories impacts people's behavior (for example, a system of superstitions) can be a contributing component of a society's culture.

2-30. Value systems differentiate those feelings, thoughts, and behaviors considered *right* from those considered *wrong*. Value systems often develop from larger belief systems. In the United States, for example, the value system behind the Good Samaritan Law (which is designed to protect citizens from being sued when they help in an emergency) is a direct descendant of the Christian belief system.

2-31. Religion plays a fundamental role in the development of a society's values. Careful analysis of a region's religious systems is crucial to understanding the local value system. Religious doctrine, stories, and myths provide enormous insight into the culture of the community. Typically, these religious elements explain why life is the way it is and suggest the values that the community holds dear. Ultimately, the values which are most prized within a system of religious beliefs affect that culture's systems of social organization, their learned behaviors, and their relationship to the physical environment.

SYSTEMS OF SOCIAL ORGANIZATION AS AN ELEMENT OF CULTURE

2-32. One common characteristic of human societies is that they tend to become increasingly organized over time. Initially, most societies (or elements of society) exist as small-scale systems, or microsystems. Such organizations might include family groups, work teams, or communal groups that share tasks and products. As societies become larger and more advanced large-scale systems (or macrosystems) of social organization usually develop.

2-33. Systems of social organization grow out of their host culture and simultaneously change the host culture by becoming deeply embedded within it. The organizing systems that a society devises (or has imposed upon it) become a part of the system of cultural meanings for that society. For example, in U.S. culture, the basic premise is that all people are created equal and can advance according to their own merits. The U.S. economic system then allots a certain value to each person's productive role; therefore, citizens often are judged by the outward tokens of their advancement and value (such as houses, cars, clothes, and leisure pursuits). The organizational systems are, in this sense, inseparable from the cultural meaning systems. One cannot fully understand one without understanding the other.

FAMILY

2-34. The term "family" refers to groups of individuals that are linked by blood-relation. In a less formal way, it may also be used to indicate a group of persons who live and work together in order to satisfy basic collective needs and goals. The role of the family is to care for and educate offspring. The family provides for such needs as food, shelter, and clothing, and prepares offspring to be functional members of the society in which they live.

2-35. The concept of family is heavily steeped in religious and other cultural norms. The family is one of the most influential elements of social institutions. It is the starting point for the development of all other forms of social organization. Families teach human beings what is demanded of them in the real world; it is within the family unit where humans learn to transcend themselves. A person's first experiences with power and authority occur with the family.

2-36. It is important for the SF advisor to remember that counterparts are equally affected by the family, although it may be in very in different ways. Even if counterparts may have similar cultural values and norms, they will act upon those values and norms differently—each in accordance with differing family values.

RELIGION

2-37. Religion is more than just a belief in a deity; it is a philosophy and a way of life. Religion can define who people are, how they view the world around them, and how they interact. In every society, people have searched for "the meaning of life." The need to understand why things happen and what will happen in the future is an ongoing epic that continues to shape people's beliefs and values.

Religion as a Component of Culture

2-38. Religion is a component of culture. At its most basic level, religion is both the individual and communal expression of contact with supernatural forces. Although it is common for individual practice to be considered important, most religious people also choose to meet with others of similar beliefs. Expressing a common religious understanding helps people to make sense of the events and issues of their world by providing explanations for human suffering, natural disasters, broken relationships, inequality between mankind, and death. Religions provide a perspective that describes life beyond the grave.

2-39. All societies have some variety of religion—typically a set of sacred beliefs and rituals that control the members by providing a common understanding of moral codes and proper conduct. Not all religions have a supernatural basis. The Marxism-Leninism ideology is often cited as a secular belief that has all the salient characteristics of religion, including the demand for faith. Similarly, some contemporary forms of environmentalism are nearly indistinguishable from ancient forms of earth worship and incorporate forms of priesthood and the recognition of a godlike Earth.

2-40. If a society is insulated for long enough, the beliefs and customs tend to become harmonious and interdependent. However, these concepts frequently appear to lose their influence when changes occur rapidly and frequently. For example, a contemporary social problem in Western society is the breakdown of common understandings—particularly moral understandings—brought about by the last 50 years of rapid societal transformation. In other cases, these beliefs may become stronger, more fundamental, and more extreme in the face of significant change, as in the case of segments of Islamic and Hindu societies in the past 30 years. These beliefs may then clash with more modern and secular approaches.

2-41. In simple agricultural communities, such as the aboriginal societies of the northern Philippines, everyone tends to do a part of all essential activities and everyone tends to have similar views of life. The activities and behavior of any one man or woman is indistinguishable from any other man or woman in the community. In such societies, each man performs the same rituals for the security of the crops and for a sense of inner wellbeing. The concept of gods and the notions of good and bad conduct are very much the same for everyone in the community. An advisor who learns about any one member of such a society learns a great deal about every member that entire society.

2-42. In complex societies the division of labor is high. No single person does more than a small part of necessary tasks. The people who participate in this division of labor are not homogeneous (as is the case with more self-sufficient, primitive societies). Because no single person understands the entire process, individuals rely upon one another to accomplish tasks. The ideas and understandings of any one member of an advanced society do not have the "completeness" that is typical of a more self-sufficient society.

2-43. Because religion is such an integral part of culture, careful mission preparation and analysis should examine the religions and religious groups in the area of operations (AO) for a given mission. Most of the people of our world practice religion and many take it very seriously. Religious beliefs, leaders, and institutions are central to the worldview of many societies. The impact of religion on the local population must be considered when planning any operation.

Common Religious Themes

2-44. For purposes of this TC, it is impractical—if not impossible—to include an in-depth analysis of every world religion. Although each religion has different beliefs, number of adherents, and spheres of influence, most religions share six common dimensions. The six dimensions are—

- Doctrine.
- Myth.

- Ethics.
- Ritual.
- Experience.
- Social organization.

Doctrine

2-45. Religious doctrine is the collection of teachings of a faith tradition. Doctrine permits religious groups to reference their beliefs and pass them on to others. It provides a body of teaching that can be communicated to successive generations. The monotheistic traditions (such as Judaism, Christianity, and Islam) are replete with doctrine. These faiths utilize sacred texts not only to record their doctrine, but also as primary bases of authority. Other belief systems, such as animism, have little if any formal doctrine.

Myth

2-46. It is important to understand that in the academic field of religious studies, the term myth carries no overtone of falsehood. The religious connotation of myths describes the narrative stories of a faith tradition that capture the truths of common belief among believers. Most religions have myths that share the essence of their beliefs with others. Although many religious myths appear primitive, they can be very effective tools in passing on the primary content of faith traditions. Such stories serve as effective vehicles to remember truths and events.

Ethics

2-47. Moral codes and ethics provide faith traditions with justification of their prohibitions of certain behaviors or beliefs. All groups need boundaries to govern their participants; moral codes provide these boundaries within religion groups. The term ethics describes the deliberate justification of how a group labels actions as either moral or immoral.

Ritual

2-48. Most religions make sacred certain spaces, times, persons, things, and events in worship. In this context, "sacred" means holy, separate, or special. Religions attempt to affect transactions between humans and the supernatural forces. Rituals are the evidence of these transactions and represent the collective and accepted ways of approaching this interaction. They are repeated in sequence and kind for continuity. Sacred persons and symbols have power and value within faith traditions.

Experience

2-49. Religious individuals and groups report that their traditions provide vivid and lasting experiences with the supernatural forces. Participants believe with conviction that they experience the supernatural and also transcend this world's limitations to reach heights of knowledge, bliss, insight, and understanding. Many traditions offer trancelike escapes from the problems of the world and allow individuals to experience a temporary exposure to realms beyond this world.

Social Organization

2-50. Religions provide individuals an understanding or explanation of the supernatural. Religious traditions unite individuals into groups and institutionalize their collective faith in social forms (for example, organizations or congregations with established hierarchies).

Broad Impact of Religion

2-51. In almost every society, the followers of the dominant religious traditions have a strong societal impact. This influence may be exerted at varying degrees upon the individual citizen, sections of society, the economy, the military, and the political structure and environment.

Individual Impact

2-52. Religions impact individuals by addressing age-old questions about identity and purpose. These questions of identity and purpose provide challenges to an individual's perspective. Religions provide the framework to answer these questions and provide personal moral codes for behavior. Some religions may provide a sense of hope, while others impart feelings of resignation.

Social Impact

2-53. Religions help to define community for subgroups of larger societies. The belief systems of religious groups provide the normative codes of conduct for group members. Religions can serve to legitimize or disqualify leaders of society based on their practices and personal lifestyle. Religions often serve as the primary collectors and maintainers of scholarship. Capable of weathering social change, many religions and their institutions provide a degree of stability in the midst of shifting influence among other groups.

Economic Impact

2-54. Practically every religion addresses the means in which their adherents should acquire, use, and distribute their resources. Participants are challenged with an obligation to care for the less fortunate and to support the collective effort of spreading their beliefs. Religions impact taxation, banking, and employment practices by dictating acceptable and prohibited forms of work and levels of profit. Religious tenets are considered in the formation of international agreements and often influence foreign economic policy.

Military Impact

2-55. Religions may describe acceptable military conflicts, military service, and the treatment of noncombatants and enemy prisoners of war (EPWs). Religions enable groups to view their enemies as evil and their conflicts as mandated by a divine force. Many conflicts arise when groups are not permitted to practice their beliefs without restriction. Religions help to define the reasons why and when certain conflicts are deemed acceptable (for example, Christianity's "Just War" theory), and how these conflicts should and should not be fought.

Political Impact

2-56. Religions may influence the rise or fall of political leaders, policies, and issues. Even in secular systems where the separation of church and state are widely accepted and enforced, religious groups often wield tremendous political power. Many societies struggle to balance the desire to permit religious expression with the unwillingness to promote every religious group as valid. Political elites may practice a religion that is different from the majority of the population. As such, leaders often face scrutiny from religious groups for their beliefs and practices. In some nation-states, religious groups form political parties and are awarded seats of representation in the parliamentary governmental bodies based on number of adherents.

The Study of Religion for Operational Purposes

2-57. SF Soldiers should approach the analysis of religious groups in an AO by remaining objective and keeping an open mind. Soldiers may draw from their experience, but they should avoid judgmental conclusions that reveal an attitude of superiority. A thorough analysis should address the interaction of religious groups within and beyond the AO and the potential impact of these groups upon the mission and the force.

2-58. Religious area studies should begin with a review of the history of the predominant religions in the region and the AO. Planners should attempt to trace the growth, influence, and changes that religious groups made over time. As a general rule, the following periods of history should be analyzed:

- Ancient through the industrial eras (up to 1900).
- World War I (WWI) (1900–1930).
- WWII (1930–1950).

- Recent decades (1960–1990).
- Present day.

2-59. A careful analysis should address how religious groups weathered social movements, global conflicts, and the postcolonial creation of nation-states. It is often possible to categorize the "winners" and "losers" of a conflict by their religious groups. Because religious conflict with a violent dimension is typically virulent and long-lasting, planners must analyze the historical perspective to fully understand the positions taken by religious groups and leaders. Additional factors that should be addressed in a religious area study are described in Figure 2-9, pages 2-14 and 2-15.

History or Background	Attempt to determine which religions are truly indigenous and which were introduced to the AO by peaceful migration of traders, by conflict or conquest, or by intentional missionary efforts. Attempt to track how religious groups have weathered the changes in the region.
Leadership	Determine how the leaders of a religious group are selected, trained, ordained, rewarded, and disciplined. The centers of learning, bases of support, and missionary efforts of religious groups provide important clues about their political and social agendas. Charismatic leaders have caused groups to revolt and act on religious impulse to conduct acts of terrorism.
Organization	Determine the levels of hierarchy for the religious groups. Describe the links between leaders and followers, leaders and other leaders, and groups and subgroups. Identify if these links are formal or informal. Determine if cell groups meet without direction or if the meetings are controlled and scheduled. Identify the ties between the centers of learning and those in positions of power for religious groups. Establish the chain of command for religious leaders, particularly those involved in negotiations or making pronouncements.
Response to Society	Determine how the religious groups in the area respond to society. Sociologist Max Weber describes two types of responses to society that religious groups adopt: control or withdrawal. Some groups may choose to control the society to which they belong. These responses may include religious movements, secret societies, social protest movements, and political parties. Some groups attempt to withdraw from the surrounding society. These responses include symbolic separation (that is, a subculture) and intentional segregation (for example, a commune). A group's theology or beliefs may dictate if they select responses that call for active resistance or passive reform.
Response to Minority Groups	Identify how religious groups interact with minority groups in the AO. Ted Robert Gurr identifies four major types of societal responses to minority groups— containment, assimilation, pluralism, and power sharing. These categories describe the varied attempts by those in power to pacify, neutralize, or divide minority groups within society.
Sites and Shrines	Identify the places of worship, sites of pilgrimage, memorial or commemoration sites, cemeteries, and other locations of veneration. These buildings, statues, and shrines may be listed on a preclusion list (in accordance with [IAW] the Law of Land Warfare). The list also keeps U.S. forces informed of possible locations of rallies, paths of pilgrimage or migration, and sensitive areas where maximum psychological effect might be achieved with an attack by enemy forces. Because SF may be the only U.S. personnel in the area, such listings may be extremely useful.

Figure 2-9. Religious area study recommended topics

Calendar	Note the normal and regular days of worship or observance. Identify special holy days of festival, feasts or fasts, celebrations, or services. Note which festivals and observances suspend normal activity. Determine special anniversaries that mark religious conquests, defeats, or reconciliations between groups and parties. Observe and respect the different calendars used by different religions (for example, describe start times for operations and negotiations using several calendar dates).
Tolerance	Determine how tolerant group leaders and members are of other groups operating in their base of support, members who exhibit bad behavior, and conversion of members to other traditions. Describe how difficult it is to join or quit the group. Identify if the beliefs of the group reinforce tolerance or exclusion toward those that differ. Determine the impact of individual conversion or initiation on the family unit, especially if others choose not to join.

Figure 2-9. Religious area study recommended topics (continued)

THE STATE: A POLITICAL SYSTEM WITHIN THE CULTURAL SYSTEM

2-60. When cultures evolve into civilizations, one of the systems of social organization that typically develops and grows in complexity is government. Throughout history, formal government (or the "state") is most often associated with urban civilizations where the economy supported numerous specializations. City growth resulted from sustainable agriculture and technological development, and, as cities developed, social hierarchies emerged with identifiable elites. The increasing complexity of life in large communities required a state organization.

2-61. States are formed as a result of a number of factors working together. For example, as population grows in a specific region, competition for space and resources also grows. Eventually, the need arises to create more complex organizations to govern effectively. The groups that organize most effectively to improve or to defend themselves gain an important advantage over their neighbors. Another common contributing factor is the specific challenges or opportunities presented by the environment. For example, a dry but fertile flood plain might support wide-scale food production when properly irrigated, and a large irrigation project is best done by a state.

2-62. The ruling class must establish that the rule of the group or individual in power is beneficial, right, and necessary (or unavoidable). Rulers often attempt to establish a cultural association between the leadership position and the values of the dominant local religion.

2-63. Unlike the Federal Government of the United States, many countries have only a single central government, and all governmental entities are extensions of it. Generally, such governments have the three following levels:

- A national government.
- Several regional entities (for example, states, cantons, districts, sectors, or provinces).
- Numerous municipalities.

2-64. Regional and municipal levels of government have no independent authority and may not, for example, levy taxes or establish budgets. Each level is an administrative extension of the next higher level. The fact that members of local government are locally elected does not make them independent of the national central government, which exercises its authority through the sector or province.

2-65. Likewise, military authority may or may not be exercised directly by the central government. In some cases the midlevel governmental chief or even the senior municipal authority may have military command authority. In other instances the military is largely independent of civilian control at any level.

CULTURAL INDOCTRINATION

2-66. The value of cultural indoctrination is best introduced through a sample of quotes from anonymous SF participants in various operations. They emphasize the value of cultural indoctrination and sensitivity training in SO and especially counterpart relations.

Cultural Indoctrination

"You have to learn about the country's history, the customs, the way of life... things that make the country tick... then you can deal with them. You can't go in there and expect them to do things your way. If you show you understand them, respect their ways, and treat them fairly, they'll be easy to negotiate with. You'll win their trust."

"You need to get a feel for the native sensibilities to be functionally effective in a community. You need to understand the nuances. You don't have to be sympathetic in terms of being politically inclined to their view, but you have to be able to absorb why they are fighting."

"Honor and 'face' mean everything here. The volume and tone of your voice are important... you could 'lose face' quickly if you lost your patience with an interpreter or got overly emotional about anything. The respected Buddhist way was to be calm, gentle, a low tone in your voice."

2-67. Culture determines a person's perception of the world. The manner in which each culture views its surroundings may differ greatly from place to place. Areas that tend to have the greatest variation of cultural viewpoints include—

- Family.
- Gender roles.
- Individualism (versus group emphasis).
- Age.
- Friends.
- Status.
- Hygiene.
- Personal space.
- Time.
- Education.
- Gestures.

2-68. In most cultures, ideology leads individuals to see an ideal version of their own culture rather than the one that really exists. For example, there is a very large gap between the American culture perceived by U.S. residents and the perception held by the rest of the world. Likewise, as foreigners, SF Soldiers may view the host-nation (HN) culture through the prism of their own idealized culture. This creates a large gap between the culture SF Soldier sees and the HN's real culture.

PROGRAM OF CULTURAL INDOCTRINATION

2-69. The general theme of cultural indoctrination should be to gain the trust of the local people by demonstrating an understanding and sensitivity towards their culture. A summary of training recommendations is provided in Figure 2-10, page 2-17.

2-70. An awareness of the cultural aspects of the AO can significantly enhance the effectiveness of an advisor. An extensive study program is best, but even a concise program is worthwhile if it provides a cultural indoctrination. The best programs involve credible speakers with personal experience in the country. Also, the training process should be dynamic and move beyond the standard lecture-with-handout format. For example, small groups or a discussion panel are most effective. Cultural behavior and cultural familiarization handbooks can be extremely useful. Because such handbooks vary greatly in accuracy and quality, subject-matter experts should be consulted to determine which handbooks to incorporate into the cultural study program.

Training Recommendations	
Cultural Indoctrination	• Comparison of cultural values and social structures (United States compared to those of the AO). • Local customs and traditions (for example, greetings and dos and don'ts). • Geopolitical history (precolonial to contemporary, and the orientation of each faction or party). • The role of religion in daily life.
Cultural Awareness	• How to gain acceptance and trust. • How to maintain a neutral perspective (for example, avoiding stereotyping and being aware of bias). • How to gain cooperation during investigations and information-gathering sessions. • How to avoid embarrassing or potentially dangerous situations.
Training Tools	
Resources	• Guest speakers native to the country of interest (for example, NGO staff, foreign students, recent immigrants, or selected refugees). • Others who have worked in or studied the mission area (for example, SF personnel, diplomats, and scholars). • Cultural familiarization handbooks.
Format	• Combination of briefings, small group discussions, and question and answer periods. • Handouts to augment—not replace—speakers. • Visual media, specifically slides and videos of the mission area.

Figure 2-10. Cultural training recommendation

2-71. The vast majority of those with operational experience stress the importance of cultural indoctrination training. Many civilian U.S. agencies (such as the United States Agency for International Development [USAID]) have developed intercultural effectiveness programs designed for their personnel working overseas. Often these programs can be revised and tailored for military and paramilitary groups. If available, these programs are an excellent resource for predeployment preparation.

Note: It is possible for Soldiers to try so hard to absorb every detail about a new culture or country that they will actually suffer from effects similar to culture shock, even though the individual has not yet left their home station. Cultural training should be thorough but not overwhelming.

CULTURAL AWARENESS

2-72. The cross-cultural communications capabilities required to perform as an effective advisor to a foreign counterpart can be described in three levels. At the lowest level is awareness, followed by knowledge. The highest level is reached when these two are combined with well-trained and refined skills.

Awareness

2-73. Awareness is the basic level of cross-cultural capability. Awareness of cultural differences and their impact is the first prerequisite for successful work with a counterpart. Simply being sensitive to the fact that differences exist and carefully observing actions and reactions can assist the SF Soldier in adjusting behavior and modifying actions to achieve greater influence with the counterpart. Awareness is not region-specific, and can be instilled in SF Soldiers with relatively little training.

Knowledge

2-74. Knowledge of the details and nuances of a specific target culture is the next level of cross-cultural capability. This second level is attained through a combination of academic study and immersion. Such knowledge is inherently area-specific and does not transfer from one target area (or culture) to another. Developing the in-depth area or regional knowledge necessary for effective cross-cultural communication requires an extensive and time-consuming training regimen and should be supported by appropriate personnel assignment policies.

Skills

2-75. Skills fundamental to effective cross-cultural communications, when combined with awareness and knowledge, form the highest level of cross-cultural capability. Although some individuals show greater natural talent for these skills than others, all SF Soldiers require continual training in each in order to achieve and retain their full potential as advisors to foreign counterparts. These skills are—

- Professional competence.
- Language.
- Nonverbal communication.
- Negotiation.
- Interpersonal skills.
- Observation.
- Problem solving.
- Leadership.
- Instructional techniques.
- Fitness.
- Region-specific skills.

Professional Competence

2-76. Given the nature of the SF missions, professional competence is critical to personal credibility. Without a high degree of credibility, advice is likely to be disregarded. Demonstrated professional competence in one area leads to a presumption of competence in other areas, including some that may not be directly related. Additionally, operational experience—particularly successful combat experience—is extremely useful, as this is universally considered a hallmark of professional military competence. As with all experience, however, SF Soldiers must be careful not to overstate the applicability of their particular experience—foreign counterparts may consider this simple bragging. SF Soldiers, as the Army's "Quiet Professionals," must take a balanced approach to conveying experience to counterparts. Unit training programs should stress the following subjects and assist in relating them to the SF Soldier's target region and culture:

- Technical and tactical expertise.
- A thorough knowledge of command and staff processes.
- Sound theory of conventional and unconventional warfare.
- Solid basic soldier skills (for example, marksmanship and map reading).
- Personal fitness.

Language

2-77. Language is fundamental to cross-cultural communications. The greater the proficiency in the local language, the more easily, quickly, completely, and accurately the SF Soldier can communicate. Complete, timely, and accurate communications are fundamental to mission success. Reaching a level that permits the advisor or liaison to understand nuances and inferred or implied messages vastly enhances communications capabilities. Proficiency in the use of an interpreter can be substituted for language ability; however, no matter how skillfully an interpreter is employed, their use always diminishes the capability to effectively and confidently communicate. Both language proficiency and the proper use of interpreters require significant training. Appendix C discusses the use of interpreters and translators.

2-78. The perceived commonality of language may be misleading. In many situations, what an individual thinks they said is not what they actually did say. Likewise, an individual may hear something that was not said or intended to be said. Each organization has its own unique jargon and has its own interpretation of key words and phrases. SF Soldiers must learn these variances and use terms and phrases appropriately. Body language is of equal importance. What one organization perceives as poise may be interpreted by another as carelessness. Formalities accepted in one organization as normal courtesies might be viewed by other organizations as overly rigid and limiting. The SF Soldier should not automatically adopt the other organization's norms; however, they must remain cognizant of the impact those norms have upon the members' interpretations of words and actions.

Nonverbal Communication

2-79. Deliberate use of nonverbal communication (that is, gestures, posture, and positioning) can be taught. The use of nonverbal communications, together with culture-specific knowledge, enhances verbal communication and understanding. Incorrect or improper use of gestures or other nonverbal signals can inhibit communication, destroy relationships, and even result in mission failure. Although simple awareness of the impact of nonverbal communications provides some level of preparation to the SF advisor, detailed regional and cultural training in their use is essential. Because most nonverbal communication occurs spontaneously and subconsciously, repetitive practical exercises should be integral to any SF training program.

Negotiation

2-80. Typically, advisors lack institutional or positional authority. They must negotiate effectively with their counterparts (sometimes from a position of relative weakness) to persuade them to take desired actions. The skill of negotiation is dealt with extensively in Chapter 4.

Interpersonal Skills

2-81. Interpersonal skills are effective techniques used to establish cordial and mutually respectful relationships between two or more people. Such techniques are employed in face-to-face settings, whether one-on-one or in small groups. They include tact, tolerance of individual idiosyncrasies or cultural norms, conversational skills, personal hygiene, and courtesy. Training in interpersonal skills requires formal instruction but is easily integrated into training designed to accomplish other objectives.

Observation

2-82. Advisors depend on their ability to observe and interpret their environment, including their counterpart's actions and reactions. Noting details of the setting and activities framing an exchange with a counterpart (as well the nuances of the counterpart's response) can provide a degree of understanding that leads to mission success. Recollection of such details and nuances facilitates post-encounter analysis to capture what occurred and to enhance future exchanges. Other cross-cultural skills (such as negotiation) also rely heavily on the ability to observe accurately and comprehensively. Like interpersonal skills, training in the techniques of effective observation and recollection lends itself to integration into other training.

Problem Solving

2-83. Advisors must develop and employ keen problem-solving skills. Templates or solutions developed in advance are seldom adequate in dynamic and unpredictable circumstances. Doctrinally accepted U.S. tactics, techniques, and procedures (TTP) must be adapted and articulated in culturally acceptable and supportable terms. Problem-solving skills and confidence in one's own problem-solving abilities improve significantly with training.

Leadership

2-84. Basic leadership skills, when properly adapted to the counterpart's culture, are very effective. Emphasis, however, should be on peer leadership techniques. Leadership techniques based on positional authority or other forms of coercion are of limited use to advisors and generally have long-term negative effects. SF Soldiers must be trained to employ leadership techniques appropriate to their role as advisors and to specific regions and cultures.

Instructional Techniques

2-85. Advisors must be competent in both formal and informal methods of instruction. Once again, these methods must be carefully analyzed and adapted based on cultural norms and practices. An emphasis on informal methods of instruction—particularly those that are (or appear to be) cooperative—are most effective in cross-cultural communications.

Fitness

2-86. Physical skills and the resulting personal fitness can be critical to personal and professional credibility when dealing with military counterparts. Fatigue reduces overall capability, diminishes perceptiveness, impairs thought processes, and impedes cross-cultural communication. Physical training programs are readily adaptable to the requirements of a given target culture and region.

Region-Specific Skills

2-87. As knowledge of a given culture is developed, certain activities or capabilities may emerge as critical to cross-cultural communications and effective counterpart relations. For example, some level of proficiency in a popular local sport may help establish interpersonal relationships that permit more effective communication and greater influence. Soccer is an example of widely played sport. More localized examples include horseback or camel riding, swimming, trapping, or even singing, dancing, and playing musical instruments. Small investments in skills seemingly unrelated to military mission requirements can provide disproportionately high returns in influence.

Operational Planning Assistance and Training Teams in El Salvador

The operational planning and assistance training team (OPATT) mission required less than half of the 55 trainers permitted in country—generally fewer than 20. With such a small number of advisors allowed for the mission, and with six brigades to support, it was critical to ensure that only qualified officers and noncommissioned officers (NCOs) were assigned to the teams. This was relatively easy with the O&I [operations and intelligence] ETSS [extended training service specialist] positions, which were filled in less than 6 months from the field of experienced senior NCOs and warrant officers, or WOs, from the 3rd Battalion, 7th SF Group.

Most SF senior NCOs and WOs had O&I education and experience, and they were capable linguists as the result of both formal training and practice on multiple training missions in the region. It was more difficult with the team chiefs. *[Continued]*

**Operational Planning Assistance and Training Teams
in El Salvador (continued)**

Not surprisingly, most officers in the initial selection, like the NCOs and WOs, had served in the 3rd Battalion, 7th SF Group. Most had extensive experience in the region from numerous mobile training teams (MTTs) in El Salvador and Honduras, and, significantly, most had worked professionally with each other in past assignments.

Cecil E. Bailey
OPATT: The US Army SF Advisors in El Salvador

CULTURE SHOCK AND ADAPTATION

2-88. The term *culture shock* was first introduced in 1958 to describe the anxiety experienced by persons in a completely new environment. This term expresses the lack of direction, the feeling of not knowing what to do or how to do things in a new environment, and not knowing what is appropriate or inappropriate. The feeling of culture shock generally sets in after the first 2 weeks of arriving in a new environment. The period of adjustment lasts approximately 6 months for most people.

2-89. Culture shock occurs because the mind and body have to go through a period of psychological and physiological adjustment when individuals move from a familiar environment to an unfamiliar one. The cues received by all of the senses suddenly change. During the day the foreigner is bombarded with unfamiliar sights, sounds, smells, tastes, languages, gestures, rules, requirements, interactions, demands, systems, and expectations.

2-90. At night—even during sleep—the brain continues to process unrecognizable sounds, the nose continues to detect unfamiliar odors, and the stomach continues to process unfamiliar foods. Even dreams seem to contain new and unfamiliar features and characters. Culture shock encompasses the cumulative effect of all of these stresses. Further psychological disorientation is brought about when values held absolute are brought into question because of cultural differences. Soldiers tend to grow particularly frustrated when they are expected to function with maximum proficiency in situations where the rules have not been adequately explained.

2-91. Over time, individuals have no choice but to adapt to their environment. There are two problems associated with this acclimation. First, the spoken rules of a culture, such as favored foods, may not be simple or pleasant to adopt. Second, the unspoken rules of a culture may be difficult to identify. Although the native members of a culture know all the rules, they may not be capable of adequately articulating those rules. Newcomers may need to be resourceful in order to extract the most basic information about why things are done in a certain way and at a certain time.

2-92. Foreign societies may also have culture-based expectations that are unknown to the newcomer. Such expectations frequently surface during first few months in a new country. It is extremely stressful for Soldiers to know that there are multiple expectations at every turn and not know exactly what those expectations are or how to fulfill them. Psychologists agree this stress is a major contributor to culture shock.

2-93. As described in preceding paragraphs, culture shock is not the result of a single event. Rather, the condition develops slowly from a series of minor events or conditions. Both the causes and the effects may be difficult to identify. Furthermore, because the human reactions are emotional, they are not easily controlled by rational management.

2-94. Some of the differences that SF Soldiers experience between their lives at their home station and their lives when deployed to a foreign location are obvious. These differences include language, climate, religion, food, educational system, and the absence of family and friends. Other differences may not be as obvious. These differences include how people make decisions, spend their leisure time, resolve conflicts, express their emotions, and use their hands, faces, and bodies to express meaning.

2-95. Being immersed in a society with extreme cultural differences tends to cause feelings of uncertainty and anxiety. The body and mind may react in unusual ways. Some persons may experience more pronounced physical symptoms of stress, such as chronic headaches or upset stomachs. Although uncomfortable, some degree of culture shock is a normal part of the adjustment process. Some common reactions include—

- Feeling irritable with (or even anger toward) one's own group or organization.
- Feeling isolated or alone.
- Tiring easily.
- Changing normal sleep patterns (either too much sleep or not enough).
- Suffering minor (but persistent) body pains, especially in the head, neck, back, and stomach.
- Experiencing feelings of hostility and contempt toward local people.
- Withdrawing from the local population (that is, spending excessive amounts of time alone reading or listening to music).

RECOGNIZING CULTURE SHOCK

2-96. There are four distinct stages in the culture shock process (Figure 2-11, page 2-23). Some analysts include a fifth stage, known as reentry shock.

Stage 1—Enthusiasm

2-97. Stage 1 is best described as the incubation period. In this stage, the new arrival may feel self-confident and pleasantly challenged. He or she may be delighted by all of the new things encountered. This is the "honeymoon" period; everything is new and exciting. Stage 1 typically lasts about 2 weeks.

Stage 2—Withdrawal

2-98. In Stage 2, a person begins to encounter difficulties and minor (but annoying) crises in daily life. It may be difficult to make oneself understood. The situation encountered may not be what was originally expected. Local nationals may prove more difficult to deal with than once anticipated. In this stage, there may be feelings of discontent, impatience, anger, sadness, and even incompetence. These symptoms occur in proportion to how different the new culture is from the culture of origin. The transition between the old methods and the new is a difficult process that takes time to complete. During this transition, there can be strong feelings of dissatisfaction.

Stage 3—Reemergence

2-99. Stage 3 is characterized by gaining some understanding of the new culture. A renewed feeling of pleasure and sense of humor may be experienced. One may begin to feel a certain psychological balance. The new arrival may not feel as isolated, and a feeling of direction emerges. The individual is more familiar with the environment and is better able to belong. This process initiates an evaluation of old ways versus new ways.

Stage 4—Achievement

2-100. In stage 4 the person realizes that the new culture has both good and bad things to offer. This stage is one of integration; the person is increasingly able to function in the new setting. A sense of accomplishment, a reduction of routine annoyances, and a more solid sense of belonging accompany this integration.

Reentry (The Fifth Stage)

2-101. Some analysts include a fifth stage—reentry shock. This occurs when the individual returns to the United States. One may find that things are not as they once were (or how they are remembered). Changes that occurred in the person's absence—their family, their friends, their communities, and themselves—combine to present a distorted image of home that differs greatly from the one imagined or remembered.

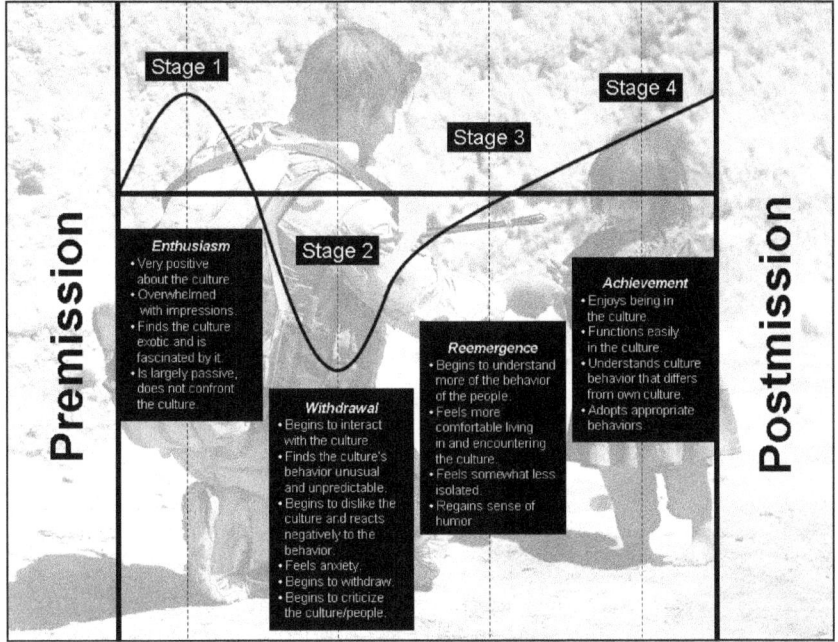

Premission

Stage 1

Stage 4

Stage 3

Stage 2

Enthusiasm
• Very positive about the culture.
• Overwhelmed with impressions.
• Finds the culture exotic and is fascinated by it.
• Is largely passive, does not confront the culture.

Withdrawal
• Begins to interact with the culture.
• Finds the culture's behavior unusual and unpredictable.
• Begins to dislike the culture and reacts negatively to the behavior.
• Feels anxiety.
• Begins to withdraw.
• Begins to criticize the culture/people.

Reemergence
• Begins to understand more of the behavior of the people.
• Feels more comfortable living in and encountering the culture.
• Feels somewhat less isolated.
• Regains sense of humor.

Achievement
• Enjoys being in the culture.
• Functions easily in the culture.
• Understands culture behavior that differs from own culture.
• Adopts appropriate behaviors.

Postmission

Figure 2-11. Stages of culture shock

CULTURE SHOCK SYMPTOMS

2-102. The stages of culture shock may be present at different times in different people, and different people react differently to each stage. As a consequence, some stages may last longer and be more difficult than others. Many factors contribute to the duration and effects of culture shock. For example, the state of mental health, the type of personality, previous experiences, socio-economic conditions, familiarity with the language, family and social support systems, and the level of education all contribute to an individual's reaction to culture shock.

Common Symptoms

2-103. Certain symptoms of culture shock are fairly common and less severe. Such typical symptoms include—

- Homesickness.
- Boredom.
- Lethargy.
- Withdrawal.
- Irritability.
- Hostility.
- Irrational anger.
- Gastrointestinal distress.
- Disruption of normal sleeping patterns.

Less Common Symptoms

2-104.   Other symptoms of culture shock present a more serious threat to the individual's physical, psychological, and emotional health and may compromise the success of the mission. Symptoms of acute culture shock mirror those produced by other forms of stress, and include—

- Sadness or depression.
- Loneliness.
- Preoccupation with health.
- Allergy-like symptoms.
- Feelings of vulnerability.
- Anger, irritability, resentment, and an unwillingness to interact with others.
- Inability to identify with any culture other than that of the United States.
- Loss of identity.
- Inability to solve simple problems.
- Loss of confidence.
- Feelings of inadequacy or insecurity.
- Obsessive behavior (for example, being overly concerned with cleanliness).
- Longing for family.
- Feelings of being lost, overlooked, exploited, or abused.

FIGHTING CULTURE SHOCK

2-105.   Although an individual may experience real pain from culture shock, it is also an opportunity for learning and for acquiring new perspectives. Culture shock can make one develop a better understanding of oneself and stimulate personal creativity.

2-106.   One of the most important tools to overcome the obstacles of a new environment is familiarity with the language. An ability to communicate in the new culture, even at the most basic level, pays enormous dividends in reducing the effect and shortening the period of adjustment. Other means of combating culture shock include—

- Having previous experience in the area. Familiarity is one of the greatest reducers of stress.
- Developing a portable hobby.
- Being patient. The process of adaptation to new situations takes time.
- Learning to be constructive. Soldiers should strive to learn from unfavorable encounters and avoid repeating them.
- Not trying too hard. Soldiers must give themselves a chance to adjust.
- Learning to include a regular form of physical activity personal routines. Exercise helps to combat stress in a constructive manner.
- Practicing relaxation and meditation. These activities have proven to be very positive for people experiencing periods of stress.
- Maintaining contact with teammates. By paying attention to the organizational relationships, Soldiers maintain a feeling of belonging and reduce feelings of loneliness and alienation.
- Maintaining contact with the new culture.
- Improving local language skills.
- Volunteering in community activities (if appropriate). Such activities allow the Soldier to practice the language and meet more people. This helps to reduce stress about language and allows the Soldier to feel more useful.
- Accepting the new culture. Time spent criticizing the culture is time wasted. Soldiers should focus on getting through the transition by thinking of (at least) one thing each day that is interesting or likeable about the new environment.
- Establishing simple goals and continuously evaluating progress.

- Finding ways to live with the things that aren't entirely satisfactory.
- Maintaining confidence in self, in the organization, and in the mission.
- Looking for help with the stress when required.
- Recognizing that uncertainty and confusion is natural. It may be helpful for Soldiers to imagine how a local resident might react to living in the United States.
- Observing how people in the new environment act in situations that are confusing. Soldiers should try to understand what they believe and why they behave as they do. Soldiers should avoid judging things as either right or wrong—they simply should be viewed as different.
- Remembering methods that have been successful in reducing stress in difficult situations in the past and applying those methods to present circumstances.
- Trying to see the humor in confusing or frustrating situations. Laughter is one of the greatest stress relievers.
- Accepting the difficult challenge of living and functioning in a new cultural setting. Soldiers must believe that they can learn the skills required to make a satisfactory transition and should gradually try to apply some of these learned skills.
- Recognizing the advantages of having lived in two different cultures. Meeting people with different cultural backgrounds can enrich life. Soldiers should attempt to share time with many different people and think of ways to help local residents learn how Americans believe and act.
- Acknowledging even slight progress in adjusting to the new culture. Many individuals have adjusted to difficult and alien environments. Soldiers should recognize that they too will make a successful adjustment to the new culture.

STRESS

2-107. A comprehensive analysis of the complex nature of stress associated with SF deployments is beyond the scope of this TC. Instead, this TC provides a synopsis of historical stress factors and experiences, a review and comparison of acute and chronic stress, and training recommendations designed to help counter the effect of stress on deployed personnel.

Sources of Stress

2-108. SF deployments are inherently stressful. Features that may individually or collectively lead to stress-related problems include—

- Working in a dangerous environment where an atmosphere of underlying tension is common. For example, local people in the AO may have been killed or injured; law, order, and respect for human rights may be minimal; and hostile attitudes by local people may pervade.
- Not being deployed or serving as a formed unit. For example, some teams may deploy as a small group of individuals from various units. Once in theater, these contingents may be further divided and deployed to different regions.
- Being immersed in a foreign culture. Adapting to cultural differences requires Soldiers to adjust not only to the culture of the mission area, but also to the varied foreign nationals working with the unit. This can include allied or coalition personnel not from the area.
- Facing challenging deployment of short duration outside one's normal experience and terms of reference. SF personnel are expected to be effective within a short time of arriving in theater. The need to adapt and learn quickly is a major source of stress.

Acute Stress

2-109. Acute stress is experienced when an individual is involved with or witnesses unexpected, abnormal situations or incidents that are outside the realm of one's normal experience. Depending on the individual, a varying range of emotions accompanies these types of experiences (for example, anger, frustration, or guilt). Critical incidents of this nature are not common; however, certain individuals experience symptoms of acute stress. Psychological first-aid (that is, individual and group approaches to dealing with mission stressors) is fundamental to initial adaptation and effective ongoing performance. Common triggers of acute stress include—

- Investigating massacres where women and children had been murdered.
- Tending to emergencies and accidents involving serious injuries.
- Being exposed to mutilated bodies.

Chronic Stress

2-110.   Chronic stress describes the result of experiencing cumulative, day-to-day stressors from a range of sources. Stressors may be environmental, such as a Soldier attempting to adjust to extreme heat, a foreign diet, or different cultural norms. Stress may result from restricted authority, such as when a Soldier lacks any power of arrest where laws and international moratoriums are blatantly disregarded. Soldiers tend to feel stress when working with bureaucracies that are burdened with inefficient procedures and crippled by the lack of motivated, mission-oriented personnel. Other chronic stressors may be organizational, such as being forced to work on small, isolated patrol elements with members who are not team players, who lack commitment to the mission, or who disrespect the local people.

Stress Management

2-111.   Leaders must recognize the unique potential sources of stress confronting SF personnel. Mission-specific stressors must also be addressed. Once the sources of stress are identified, leaders must develop a knowledge of individual and group signs and symptoms for both chronic (cumulative) and acute (critical incident) stress. After thoroughly understanding the sources and symptoms of common stressors, leaders must strive to understand individual and team coping strategies and techniques. Common stress-reduction techniques include post-patrol team discussions and after-action stress debriefings.

2-112.   Effective stress management programs enable commanders to more effectively care for Soldiers while accomplishing the mission. The desired end state of stress management includes the following:

- Operational effectiveness is enhanced due to improved ability to cope with various in-theater stressors.
- Individual and group stressors are recognized and addressed at an early stage.
- Healthier and more realistic attitudes toward stress are developed.
- Cohesion and morale are improved when the team is viewed as a support net by its members.

2-113.   Stress management is a skill that must be trained. There are numerous techniques to develop and enhance group and individual coping techniques. Panel discussions in which personnel recount stressful aspects of previous deployments and how they were addressed can be particularly effective. A panel facilitator with expertise in stress education can guide the discussion and highlight key teaching points. Briefings that address scenarios typical of the mission area can also be valuable tools.

Note: Commanders should strive to have unit personnel with recent experience discuss stressful scenarios and coping techniques. The camaraderie felt among members of the same unit contributes to a more credible and dynamic training process.

CROSS-CULTURAL COMMUNICATION

2-114.   Communication is the transfer of messages from one person to the next. These messages may be passed along verbally, in writing, or by signals. In essence, the sender encodes the message and the receiver decodes it. The type and style of encoding used is based upon the sender's history, beliefs, values, prejudices, attitudes, and preferences.

2-115.   The receiver decodes messages based upon lifestyle, group membership and rank, worldview, status, language, and social practice. Communication is a two-way process in which the encoding and decoding methods can affect both sending and receiving.

2-116. Effective communication occurs when the message is perceived and responded to in the manner that the sender intended. Ineffective communication primarily occurs from poorly chosen words, flawed timing, a confused mixture of verbal and nonverbal signals, poor listening skills, and communication noise. Communication noise describes the influences that detract from effective communication

BRIDGING THE CULTURAL GAP

2-117. Each culture has its own rules regarding who a person may speak to, how and when the person may speak, and what topics the person may speak about. Many cultures rely heavily on nonverbal signals to communicate. In such cultures, posture, expression, and actions convey more than spoken or written words.

2-118. Language is the ultimate communication barrier. The successful advisor should study, at a minimum, common phrases in the HN's language. Additionally, advisors must recognize that there are topics of conversation that should be avoided whenever possible (for example, religion, politics, ideology, and personal questions).

Developing a Sense of Cultural Awareness

2-119. Advisors must recognize they are a product of their own culture. They must learn as much as possible about the culture of the people with whom they need to communicate. When communicating with people across cultures, advisors must abandon any sentiments of ethnocentrism—the tendency of individuals to judge all other groups according to their own group's standards, behaviors, and customs. Such notions lead an individual to see other groups as inferior by comparison.

Recognizing Differences

2-120. Each culture has its own way of accomplishing required daily tasks. Advisors must understand that each particular society may approach things differently; it does not mean that they are inefficient or less intelligent. Being different should not be seen as negative. Respect for counterparts must be maintained at all times.

Learning to Adapt

2-121. Advisors must be flexible and ready to adapt and adjust their behavior; however, they must be careful not to overdo their adjustment. Individuals who are overly flexible are often perceived as being insincere. The successful advisor must strive to act in a way that is appropriate to the target culture. Above all, advisors should be themselves and show sincerity.

Developing Tolerance

2-122. Advisors must develop a tolerance for deviations from accepted norms. Events or activities that may seem extraordinary to newcomers may be common practice in the culture. Advisors must be aware that members of the foreign culture may be astounded by that which is commonplace in the United States. Careful observation should be made before judgments are rendered about seemingly peculiar behavior.

PERSONAL CONTACT

2-123. Personal contact is the most effective way to bridge organizational barriers. Interorganizational stumbling blocks are very real and the prejudices that arise from them are exacerbated by misunderstanding and ignorance. Often it is too easy to attribute negative attitudes and hostile motives to faceless groups. The SF Soldier must employ superior interpersonal skills and deal directly and closely with individual members of other organizations. This type of contact can effectively reinforce commonality and diminish the impact of disparity. These personal relationships are the key to effective interorganizational relationships. Appendix A discusses useful techniques to establish and foster personal contact.

FACTORS AFFECTING COMMUNICATION ACROSS CULTURES

2-124.   There are five key aspects that greatly impact communication across cultures. These factors are—

- Level of formality.
- Level of directness and explicitness.
- Perception of time.
- Perception of the individual versus the group.
- Show of emotion.

Level of Formality

2-125.   Levels of formality vary greatly from culture to culture. Most Asian cultures are on the high end of the formality continuum. In contrast, the North American culture places very little emphasis on formality. Advisors must recognize the importance placed on formality and adjust their behavior accordingly.

2-126.   To most Germans, chewing gum when receiving a briefing may indicate that the individual is not paying attention. This translates into lack of respect for the speaker, and may even be construed as offensive. Conversely, Americans can be quite casual at work. Chewing gum during a presentation may not indicate lack of attention. In certain settings, it may not even be considered a lack of manners.

2-127.   In order to be successful, advisors must be aware of such cultural inconsistencies. A simple action that is completely acceptable in one culture may be considered distasteful in another. Time spent researching such issues of formality pays great dividends when serving in an advisory capacity.

Level of Directness and Explicitness

2-128.   Certain cultures are very direct and explicit in their communication, whereas other cultures are indirect and vague when expressing themselves. Most people from Asian and Middle Eastern cultures place a high reliance on shared experience, nonverbal cues, and the context in which the communication takes place. Consequently, they often seem indirect and vague in their verbal communication. In countries such as the United States, Switzerland, and Germany, people tend to rely heavily on the spoken word for communication. As a result, verbal interactions are very direct, precise, and explicit. Reliance on context here is low, as is reliance on nonverbal cues. To outsiders, this style of communication may be perceived as too direct and overly talkative.

2-129.   The indirectness that characterizes communication in some cultures often is a strategy designed to prevent another person from losing face. It may be viewed as consideration for another person's sense of dignity. In cultures that are direct and explicit in their communication, however, this indirectness may be perceived as dishonesty or insincerity, suggesting that the speaker may have something to hide.

2-130.   When communicating with people from a more indirect culture, it is important to exercise extra caution—both in what is said and how it is said. Being overly direct may result in unintentional offense. Careful attention must also be given to nonverbal cues, shared experience, and the circumstances in which the communication takes place. The true meanings of messages may actually reside more in these nonverbal cues than in the words uttered. On the other hand, when communicating with people from a more direct culture, typical American candor is more appropriate.

2-131.   When dealing with more direct cultures, ill-prepared advisors may become offended when ideas or opinions are attacked with an unaccustomed degree of directness. In order to be successful, advisors must bear in mind that some cultures employ such directness as a technique to achieve clarity. Advisors must be objective in listening to what their counterparts have to say and remember that attacks on ideas are not intended to be personal or deliberately embarrassing. Finally, because direct cultures place little reliance on context, advisors must pay particular attention to the spoken word.

Perception of Time

2-132.   Cultures may differ greatly in their perception of time. Edward Hall, a prominent researcher in the field of intercultural communication, made a useful distinction between monochronic-time and polychronic-time cultures. In monochronic-time cultures (most Western countries), members place a great emphasis on schedules, precise reckoning of time, and promptness. In such cultures, the schedule takes precedence over the interpersonal relation. Furthermore, because of this urgency to maintain schedules, members of such cultures tend to get to the point quickly. This directness may be viewed as rude or brash in polychronic-time cultures.

2-133. In polychronic-time cultures, time is viewed as fluid. Members of polychronic-time societies do not observe strict schedules—agendas are subordinate to interpersonal relations. Most Asian countries (as well as a number of Latin American and Middle Eastern countries) are considered polychronic-time cultures.

2-134. By knowing how the concept of time of embraced, advisors can better adapt to a new cultural environment. Because the United States is a monochronic-time culture, Americans must learn to be patient with polychronic-time cultures for whom punctuality is at best unimportant and at worst a negative trait.

Perception of the Individual Versus the Group

2-135. Cultures may be individualistic or collective in their orientation. Geert Hofstede, another prominent researcher in intercultural communication, describes an individualist culture as one in which the ties between individuals are loose—where people are expected to take care of themselves and their immediate families. In a collectivist culture, people are raised from birth into strong, cohesive groups. These groups offer a lifetime of protection in exchange for unquestioning loyalty.

2-136. In individualist cultures, the individual takes center stage. Independence is highly valued. Individuals earn credit or blame for the success or failure of their endeavors.

2-137. In collectivist cultures, an individual is regarded as a part of the group. Interdependence prevails among individuals in the same group. Credit or blame for success or failure belongs to the entire group. Individuals do not seek recognition and often are uncomfortable if it is offered.

Show of Emotion

2-138. Each culture has its own system of expressing emotion. Some cultures tend to be very expressive with their emotions and show their feelings plainly by laughing, grimacing, or scowling. Other cultures tend to be more repressive. Rather than showing their feelings openly, members of these cultures keep their emotions controlled and subdued.

2-139. When people from these two cultures interact, misunderstandings are common. Those from the more expressive culture may view people from the repressive culture as cold or unfeeling. Similarly, those from the more repressive culture may view their more expressive colleagues as immature and eccentric. Advisors must avoid snap judgments. In order to communicate effectively, the advisor must be able to display the appropriate level of emotion.

NONVERBAL COMMUNICATION

2-140. Nonverbal communication uses facial expressions, gestures, physical contact, and body postures to convey meaning. Although there are many excellent books on the subject, virtually every publication discusses nonverbal communication from a particular cultural vantage point. Nonverbal communication and culture are inseparable; one can only properly understand that this type of communication if one understands the culture from which it originates.

Body Language

2-141. Body language and gestures are products of custom, and they are just as important to communication as spoken or written language. The improper use of a gesture—even as simple as a thumbs-up gesture or a two-finger peace sign—can upset a carefully nurtured relationship.

2-142. The most basic survival gesture—the smile—is a visual signal that is universally understood. However, even though the smile is understood in every culture, some fine distinctions do exist between cultures. Russians, for example, very rarely smile on the street. The French believe Americans smile too much. Japanese people seldom smile during (what they consider to be) formal circumstances, such as a photograph for a driver's license or Christmas card. Malaysians and Indonesians tend to smile—even giggle—when they feel nervous or embarrassed.

2-143. The common symbol for OK, made by touching the tips of thumb and index finger, is one of the best known gestures in the United States. In other cultures, however, the same gesture may mean something entirely different. For example—

- In the south of France, this gesture is used to indicate zero or something that worthless. An individual who uses the symbol to indicate OK may negate a negotiated arrangement or offend a counterpart by indicating that the deal (or the individual) is worthless.
- In Japan, this gesture is used to indicate a coin or money. An individual who uses the symbol to indicate OK may offend the counterpart by appearing to ask for a bribe.
- In Brazil, Germany, and Russia, this gesture is used to indicate intimate body parts. An advisor carelessly using this gesture may appear to be suggesting something unintended.

2-144. Advisors must pay particular attention to seemingly minor differences. For example, Americans tend to beckon a waiter by raising a single finger. That type of pointing gesture is frowned upon in Japan, where pointing is done with the thumb of a closed fist. In most of Europe, a person beckons a waiter holding their hand in front of them (fingertips forward with the palm facing the floor) and making a scratching motion with the fingers. Another common beckoning gesture in the United States is made by holding the index finger straight up (facing inward) and curling it up and down. In Australia, Indonesia, and Mexico, this gesture is reserved for animals and prostitutes.

2-145. The use of body language in the AO should be studied thoroughly prior to deployment. Once deployed, advisors should continue to study the body language used by local nationals. The surrounding circumstances must also be considered (for example, how the people call a waiter or how they wave goodbye). Advisors should observe posture, body language, and common gestures, such as people tapping the sides of their noses, or flipping the lobes of their ears.

The Far Reach of Social Blunders

The danger of USG representatives committing cultural *faux pas* does not stop at advisor level. On one occasion, when President Richard Nixon exited an airplane in Brazil, he raised his arms and made the "OK" symbol with both hands. In Brazil, this symbol indicates an intimate body part. In 1991, when President George Herbert Walker Bush visited Australia, he offered the "V for victory" sign from the window of his limousine. His hand, however, was facing toward his body instead of away from it. In British Commonwealth countries, this gesture is the equivalent of the extended middle finger in the United States.

Interpersonal Distance

2-146. The space people maintain around their bodies reflects the desire to control which people are allowed to get close and the circumstances in which it is deemed acceptable. An individual who violates another's interpersonal space without consent is usually perceived as hostile or aggressive. The perception of appropriate interpersonal distance varies between cultures and reflects the style and tone of the society at large. Figure 2-12, page 2-31, provides an intentionally broad overview of these variations. The guidelines provided cannot be applied to every relationship or circumstance, even within a single cultural group. Few areas of nonverbal communication are more sensitive than interpersonal space. Fortunately, the appropriate distance is easy to respect once it is identified.

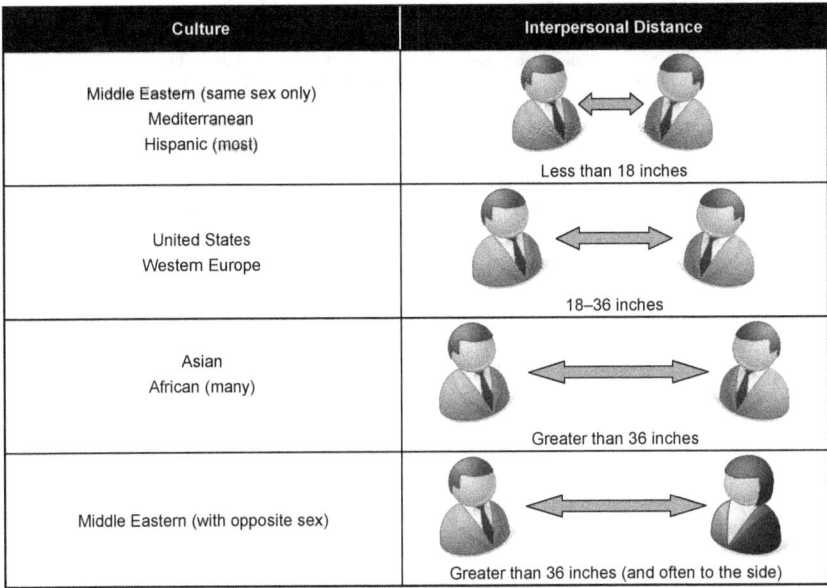

Culture	Interpersonal Distance
Middle Eastern (same sex only) Mediterranean Hispanic (most)	Less than 18 inches
United States Western Europe	18–36 inches
Asian African (many)	Greater than 36 inches
Middle Eastern (with opposite sex)	Greater than 36 inches (and often to the side)

Figure 2-12. Interpersonal distance

2-147.   For SF Soldiers serving in an advisory role, interpersonal distance is particularly important when administering or receiving criticism. In such instances, individuals must be sure they do not stand too close to one another. Too small an interpersonal gap can make one or both of the individuals feel threatened and defensive. People who feel defensive have trouble expressing themselves clearly; are resistant to suggestion, correction, and criticism; and may even become aggressive.

Touching

2-148.   Touch should be minimized when communicating across cultural lines. Although some cultures are more open to touching than others, even the most demonstrative groups have rules of propriety and etiquette. Physical contact made at the wrong time can risk serious misunderstandings.

2-149.   In mainstream American culture, touching is generally discouraged. Native-born Americans tend to abandon touch at an early age and substitute words as the primary means of communication. Northern Europeans, such as the Germans, Scandinavians, British, are similarly uncomfortable with touch from anyone other than intimate family members or friends. Touching is discouraged in most Asian cultures, particularly to the back, head, or shoulder. Additionally, small children are not to be touched by anyone outside the immediate family. Whereas it is common for an American adult to playfully tousle the hair of a small child, such actions may be deemed extremely offensive in some Asian cultures. In the example of the small child's hair, an advisor's well-meant attempt to ease tension, display a sense of humor, and build trust would achieve the exact opposite.

2-150.   There are rules for touching. Because U.S. society is so aware of the potential for people to use negative touch to intimidate or threaten, most Americans are careful in how they touch. Researchers classify Americans as low touchers in relation to the rest of the world. However, the use of touch in such a multicultural society varies. Some Americans never touch anyone outside their immediate family, even though they may prize such a person's friendship. Others touch often—usually on the shoulders and arms—but such contact is not designed to express any real meaning.

2-151.   In the U.S., touch is used mainly when individuals meet or depart (for example, a handshake or a casual hug). Americans can simulate the feeling of touch (without ever touching) by allowing others to move in close when talking. Close friends may exchange hugs, pretend punches, or kisses, and may touch frequently when talking. For acquaintances and superiors, a simple handshake is all that is expected. Some people are considered high touchers and give friendly arm, back, and shoulder touches—even to new acquaintances. Some Americans show in public what might be considered private expressions of affection in other cultures (for example, prolonged kisses or other overt displays of affection).

2-152.   Handshakes exchanged in the United States tend to be more firm than other cultures. For instance, Chinese, Japanese, and most Africans use light handshakes. This does not indicate that the individual is unassertive; it simply is a cultural trait. Russians and most Europeans use firm handshakes. Similarly, this does not necessarily imply that the individual is assertive or tough.

Eye Contact

2-153.   Rules for appropriate eye contact vary among cultures; however, few people consciously recognize that such rules even exist. Still, individuals do act IAW with society norms. These individual have unconsciously absorbed the rules of their culture.

2-154.   Among the majority of North Americans, the accepted rule is for individuals to make intermittent eye contact with the people to whom they are speaking. If people hold eye contact for too long, they may be considered aggressive. If people hold eye contact too briefly, they may come across as disinterested. An individual who consistently fails to make eye contact may appear evasive or dishonest.

2-155.   Advisors must be sensitive to cultural variations when making eye contact. In some African cultures, for example, it is considered impolite to make more than the briefest eye contact. To most Americans, such an exchange would imply that the other party was distracted and not paying attention.

Facial Expressions and Gestures

2-156.   It is not unusual for Americans to use facial expressions to convey doubt, surprise, distrust, anger, agreement, or rejection. Other cultures are less expressive with facial expressions. For example, a Japanese counterpart's face may be expressionless, but inside he may be furious. Similarly, he may enthusiastically agree with a recommendation, although his outward expression might be interpreted as indifference.

2-157.   The majority of Americans use wide hand gestures. In most European and Pacific Rim countries, hand gestures are kept to a minimum. In Latin America, the Middle East, and Africa it is common for people to use wide gestures to emphasize their points of view. Advisors must exercise caution in both the number and magnitude of hand gestures.

COMMUNICATION WITH ILLITERATES

2-158.   SF Soldiers frequently operate in developing countries—and even some industrialized nations— where large segments of the population are either illiterate or semiliterate. In many cases (such as Haiti), this segment of the population represents the primary target audience with whom the unit must communicate. An illiterate or semiliterate audience presents a challenge that requires creative and enhanced communication skills.

2-159.   In reaching an illiterate or semiliterate segment of the population, there are a number of existing channels of communication that can be used; others can be created. Radio is a powerful means of reaching illiterate or semiliterate populations. Other techniques include interpersonal communication, traditional or folk communication, simple graphics, and other basic printed materials. When dealing with illiterate or semiliterate audience, the SF advisor should—

- Determine how people receive their information. The lack of literacy skills does not necessarily indicate an absence in the flow of information.
- Identify the key communicators for the target audience. Very often, word of mouth is the best way to share information among audiences for whom written materials are ineffective.

- Identify the informal communications networks (the "grapevine") in the community. In some areas, the most effective communications networks lack leadership or formal structure. Advisors must identify members of the community who trust the unit and are willing to help.
- Include the use of printed materials. Photo-novels, comic books, and wall posters that use more graphics and fewer words can convey powerful messages.
- Cater visual communication materials to the target audience. Images that effectively convey a message in one culture might carry an entirely different meaning in another. Artwork should be tested on a sample audience.

TOPICS TO APPROACH WITH SPECIAL CAUTION

2-160. Each culture has topics that are considered forbidden—subjects that are simply not discussed in casual conversation. The nature of these topics can vary widely; however, SF advisors should avoid (at least initially) the following three subjects:

- Religion.
- Politics.
- Personal matters.

Religion

2-161. Discussing religion can immediately destroy the communication process, particularly if the local nationals believe that offense was given. Whether actual or perceived, the damage done may be very profound, as cultural sensitivities may be acute. This is evident in the case of Islam, for example.

Politics

2-162. Political discussions can become very heated and should be avoided if possible. In regions where political issues dominate the news, avoiding such discussions may prove difficult. If there is no polite way to extricate oneself from the discussion, advisors must ensure that it is handled in a diplomatic manner.

Personal Matters

2-163. It is considered rude to ask other Americans highly personal questions. Advisors should extend the same courtesy to members of the local national population. In certain rural areas of West Africa, where the world is seen in generalities, it is considered extremely rude to ask specific questions.

FINAL TIPS ON COMMUNICATION

2-164. Researching a target audience's culture can be an endless activity. Advisors must not get bogged down on the minute details of a single cultural feature. Instead, they should strive to understand all aspects of the culture. If advisors demonstrate basic knowledge and are willing to listen and learn, they will achieve a more effective exchange. Appendix D provides a number of valuable Internet resources.

2-165. Advisors should consult with individuals who lived in the HN and know the local customs. A 1-hour interview with a native may be more productive than a week of formal language and culture training.

2-166. Effective communication requires practice. Advisors should rehearse in order to practice favorable body language and eliminate unfavorable gestures and postures. Body language is as important as the verbal message. It should appear natural—not labored or uncomfortable.

2-167. Advisors must recognize that an American stereotype—possibly a negative stereotype—likely precedes them. They should try to understand and discern this stereotype. Over time, they may be able to improve the local impression of Americans. Finally, advisors must be mentally prepared to experience the unknown.

This page intentionally left blank.

Chapter 3

Dealing With Counterparts

Merriam-Webster's Collegiate Dictionary defines rapport as: "a relation marked by harmony, conformity, accord, or affinity." When people discuss good rapport, they describe a relationship that is founded upon mutual trust, understanding, and respect. Relationships characterized by personal dislike, animosity, and other forms of friction often are described as no rapport. Rapport, for purposes of the SF advisor, is a term used to describe the degree of effectiveness.

ESTABLISHING RAPPORT

3-1. The requirement to establish rapport with HN counterparts is the product of the advisor being in a unique military position. That position is one in which the advisor has no positional authority over the actions of the HN counterparts. This lack of authority means that the traditional (that is, doctrinal) view of military leadership must be modified to emphasize interpersonal relationships and deemphasize authoritarian roles. Advisors use their interpersonal skills to influence the outcome of events.

3-2. SF advisors must be armed with certain knowledge before they can establish effective rapport. First, they must study FM 6-22, *Army Leadership*. This publication provides the SF advisor with the basic leadership knowledge required to understand human nature and motivation. In addition to FM 6-22, advisors must incorporate information specific to the culture and society of their potential counterparts. This information may take the form of thorough area studies, operational area studies, and other research materials (as discussed in Chapter 2).

3-3. The true measure of rapport is whether or not the advisor can motivate their counterpart to take a desired action. For SF, the basic techniques of motivation (in the absence of authority) are advising, setting the example, compromise, and coercion.

ADVISING

3-4. Advising the counterpart to select a particular course of action is only effective if the counterpart perceives that the advisor is professionally competent enough to provide sound advice. If the counterpart does not perceive the proposed solution to a problem is realistic, the advisor's competence will be questioned. The advisor must take pains to explain to his counterpart that the recommended course of action (COA) is realistic and will be effective.

SETTING THE EXAMPLE

3-5. Setting the example for the counterpart must be an ongoing effort in order to avoid the appearance of a "do as I say, not as I do" attitude. In setting the example, the advisor should make every effort to explain to the counterpart that what he is doing is the most effective form of behavior for the situation. This is particularly true when the behavior (or purpose) is not readily understood by the counterpart. In following this guidance, the advisor may also reinforce his perceived competence.

COMPROMISE

3-6. When seeking compromise with the HN counterpart in the desired COA or behavior, the advisor may create a situation in which the counterpart has a personal interest in successful execution. In some cultures, seeking a compromise may allow the counterpart to save face. Furthermore, in certain situations the counterpart—because of practical experience—may have a better solution to the problem at hand.

3-7. Advisors must also recognize that when they seek a compromise in certain cultures, their perceived competence may suffer. This may be mitigated somewhat by approaching the compromise as two professionals (the advisor and the counterpart) reaching a mutual conclusion. In order to reach an effective compromise, SF advisors may have to conduct negotiations. The approaches and techniques of negotiations are covered in detail in Chapter 4 of this TC.

Note: There are two areas of concern that must never be compromised for the sake of maintaining rapport—force protection (FP) and human rights.

COERCION

3-8. Coercion is the least desirable method of motivation because it can cause irreparable damage to the relationship. Advisors should use coercion only in extreme circumstances (that is, life or death).

3-9. American advisors frequently enjoy a privileged status in the HN. Their mere presence may garner personal benefits for the counterpart simply because the counterpart has a one-on-one association with an American. Worried about losing these benefits, counterparts may agree with advisors simply to avoid confrontation. SF advisors must be careful to avoid unintentionally forcing their counterparts into action.

3-10. Ultimately, the SF advisor should strive to establish and maintain good rapport by conveying to his counterpart that he is sincerely interested in him, his nation, and its cause; that he will not belittle him or his efforts; and that he has no intention of taking over from him. Advisors must demonstrate that they have come to help because they believe the counterparts' goals are just, fair, and deserving of success.

POLICY AND RELATIONSHIPS

3-11. Advisors must keep in mind that their primary aim is to forward U.S. policy. Therefore, relationships with HN personnel, other U.S. forces, and OGAs must be established and developed.

SUPPORTING UNITED STATES AND HOST-NATION POLICIES

3-12. SF advisors not only support U.S. national policy in the areas where they operate, they are also obligated to forward the policies of the HN government (unless otherwise directed by higher authority). Advisors may submit periodic situation reports (SITREPs) to the Embassy and should maintain communications with higher headquarters (HQ) to ensure that activities are in keeping with U.S. objectives.

3-13. HN national policies, organization, economy, customs, and education may dictate practices and procedures that appear inefficient or uneconomical. Advisors should avoid criticizing or condemning such practices and procedures until they are thoroughly understood. Recommendations that may be critical or contrary to HN policy should be made in private.

ENVIRONMENT

3-14. The SF advisor must understand his status in the HN. This is normally specified in detail by a status-of-forces agreement (SOFA) or other arrangement between the USG and the HN government. These agreements vary widely; they may offer little political protection or may provide for full diplomatic immunity. If no such agreement exists, the SF advisor is subject to the full measure of local laws and customs and the jurisdiction of local courts. Even if a SOFA or other agreement exists that provides some form of immunity, advisors are still expected to observe local laws.

3-15. When an advisor enters a country, the HN government may still be in the process of developing adequate administrative machinery. The advisor should be aware of such scenarios and avoid being overly critical. They should remain flexible and use their creative to find approaches that work. In the absence of clear laws, a U.S. Soldier should conduct himself professionally and IAW U.S. laws and regulations and the Uniform Code of Military Justice (UCMJ).

3-16. The SF advisor must have knowledge of the political, social, and military organizations in the AO and the manner in which they interrelate. In many countries these relationships depend heavily on personal relationships between individuals. The SF advisor must understand the personalities, political movements, and social forces at work. Close contact with local civilian leaders, military commanders, and police forces helps to establish and update a working knowledge of these groups and their relationships.

COMMAND RELATIONSHIPS

3-17. Advisors must use the U.S. chain of command to obtain and disseminate guidance and assistance. Care must be taken to distinguish between the U.S. chain of command and the HN chain of command. In particular, advisors must prevent the counterpart from attempting to control subordinates through the U.S. chain of command. Advisors provide recommendations—not orders—to their counterparts. Only the counterpart should issue orders to subordinates.

3-18. OGAs, NGOs, and other HN agencies may be collocated in the AO. (Appendix D discusses these organizations in detail.) Advisors should strive to integrate the efforts of all organizations into mission planning and should impress upon their counterparts that progress can only be achieved through such integrated efforts.

3-19. Advice should be presented in person. If the advice provided is not accepted, and the SF advisor feels it is appropriate to do so, he may report the matter in writing through U.S. military channels. It is possible that HN or U.S. policy conflicts at higher levels may prevent the counterpart from acting on the advice provided. If higher echelons are made aware that a problem stems from policy conflict, they may be able to align the policies.

3-20. When reporting such matters, the SF advisor should consider that a written grievance may resurface at an inconvenient or embarrassing time or place. Such reports may have unintended and far-reaching consequences. For example, a report about an HN officer may be shared with a high-ranking HN official. The HN official, in an attempt to prove that he is working to correct the situation, may fire the officer in question.

3-21. Effective communication is essential for the advisor. The use of proper channels should be stressed at all echelons. SF advisors must keep their counterparts informed of advice given to their subordinates. Fellow U.S. personnel should be kept informed of advice offered to counterparts. HN officials should be persuaded to pass information up, down, and across the chain of command.

3-22. Local customs and courtesies should be observed. Counterparts that are senior in grade should be treated accordingly. If warranted by the HN military customs, senior-ranking personnel should be saluted. Such individuals should be referred to by their rank (as customs permit) and shown respect and deference. Although HN officers may have no command authority over the advisor, effectiveness is greatly enhanced when the advisor displays respect for the counterpart and the HN chain of command. Figure 3-1, pages 3-3 and 3-4, provides additional recommendations of common dos and don'ts for advisors interacting with their counterparts.

During Counterpart Interaction	
Do	*Do Not*
Study the counterpart's personality and attempt to influence from the background.	Attempt to command the counterpart's organization.
Make recommendations in private.	Correct the counterpart in public, particularly in front of subordinates.

Figure 3-1. Advisor-counterpart interaction guidelines

During Counterpart Interaction	
Do	*Do Not*
Represent the counterpart or defend the counterpart's position in disputes with U.S. agencies as long as it is based on sound, reasoned judgment.	Represent the counterpart or defend the counterpart's misguided position out of blind loyalty or an attempt to win favor.
Provide advice in a respectful manner.	Present suggestions in a condescending manner or in a way that might be embarrassing.
Ensure the counterpart understands the recommendation.	Provide too little, too much, or disjointed guidance.
Ensure the counterpart has the authority and capability to follow recommendations.	Attempt to force the counterpart into following guidance that he is unable or unauthorized to perform.
Verify that the counterpart understands, agrees with, and intends to implement the recommendation.	Accept "yes" at face value. (A counterpart may answer "yes" to a recommendation, meaning that he understands, but does not agree. It also may mean that he does not understand and is merely being polite.)
Present recommendations carefully and in detail, supported by sound reasoning and an explanation of the advantages they offer.	Present recommendations that require an immediate decision.
Allow counterparts to exercise their prerogative.	Make the counterpart overly dependent upon U.S. guidance or allow such a perception to develop.
Express U.S. customs with caution.	Assume behavior that is common in the United States is acceptable in the HN.
Praise the counterpart when he makes good choices.	Convey the impression that everything is all wrong.
Select words carefully and watch the message being conveyed by tone and body language.	Use sarcasm, irony, or mockery.
Be willing to ask for the counterpart's advice, particularly in matters of culture, local customs, and courtesies as well as counterinsurgency.	Present an attitude of intellectual superiority.
Present an accurate self-image and a realistic measure of capabilities.	Create unachievable expectations or promises that cannot be kept.
Develop goals and milestones and encourage inspections.	Operate without a training plan or allow training to go unsupervised.
Demonstrate and encourage initiative and inventiveness. Encourage counterpart to clarify orders and to make recommendations when appropriate. Encourage counterpart to allow his subordinates to do the same.	Develop or encourage an environment of blind loyalty.
Encourage long-term projects and maintain accurate documents for successors.	Discourage lengthy projects simply because they will not be finished during your tenure.
Participate in local military, athletic, and social functions. If unable to accept a social invitation, decline with regrets expressed IAW local custom. Invite counterparts to appropriate social functions.	Become isolated from the local community.

Figure 3-1. Advisor-counterpart interaction guidelines (continued)

CONDUCT OF OPERATIONS

3-23. SF Soldiers can expect to participate in stability operations. These are operations in which the armed and paramilitary forces, as part of the interdepartmental team, support any or all of the internal defense and development (IDAD) campaigns in a given area. The primary operational roles through which armed/paramilitary forces support IDAD campaigns and operations are:

- Tactical operations.
- CA operations.
- Intelligence operations.
- Populace and resources control (PRC).
- PSYOP.

3-24. U.S. assistance may include advice on military organization, training, operations, doctrine, and materiel. In addition, the United States may provide and control U.S. logistics and sustainment for HN military forces. The objective of this assistance is to increase the capability of HN organizations to perform their missions and operate efficiently in the given operational environment.

3-25. Organizations and individuals possessing greater skills and resources assist by imparting their knowledge through assistance efforts. The success of assistance depends largely upon effective interaction between advisors and their HN counterparts. Advisors may operate in the following capacities:

- National-level advisor.
- Operational-level advisor.
- Tactical-level advisor.

NATIONAL-LEVEL ADVISOR

3-26. The national level is considered the largest national subdivision. This may be a nation's combined military HQ, service components, national level government officials (ministries), or a combination.

Military Responsibilities

3-27. The national-level advisor provides advice to senior HN officials on matters concerning the employment of HN military and paramilitary forces under their jurisdiction. Other major responsibilities include area defense, counterinsurgency, and the procurement and employment of U.S. support. As the U.S. military representative at the national level, the SF advisor plans for and recommends the allocation of resources provided through the Military Assistance Program (MAP) and similar sources. These resources, as well as those provided by USAID and voluntary agencies, often are in support of military civic action.

3-28. The SF advisor coordinates the military civic action program with other agencies to ensure unity of effort and appropriate use of resources. USG funding may provide materiel assets while HN troops and equipment perform the labor.

3-29. SF advisors may find that operational level forces have CA and PSYOP capabilities. In such instances, the national-level advisor should assist his counterpart in planning for the proper employment of these resources.

Civil Responsibilities

3-30. National-level advisors may be the only U.S. representatives in an area and may be required to advise on civil matters. Close and continuous supervision of all IDAD programs is essential. Province chiefs (or their equivalent) likely have administrative staffs to assist in carrying out their duties. SF advisors should become familiar with the responsibilities, functions, and personnel of the administrative staff. These individuals can be a tremendous source of information.

3-31. As outside assistance to the HN increases, other U.S. personnel may be introduced into the area. At the national level, the SF advisor can expect to find representatives from governmental and nongovernmental organizations. Likewise, third-country nationals, representatives of private corporations, and local voluntary organizations may be involved in such tasks as medical care and industrial and agricultural development. Given the requirement for effective coordination, the SF advisor may find it necessary to coordinate some or all of these activities. If the SF advisor has the authority, he should see that interagency agreements are established as soon as possible. In the absence of such authority, he should actively encourage the development of such agreements. The advisor can expect to find certain technical agencies and services that are extensions of HN national ministries. Their activities and efforts should be integrated into the overall plan. This requires that the national-level advisor maintain close coordination with the USAID representative who normally has responsibility for advising these agencies.

OPERATIONAL-LEVEL ADVISOR

3-32. Operational-level advisors work at a level below the national-level advisor. This may be with military corps or division officers, with regional government officials, or a combination of all. Depending upon the size of the nation and its military, operational-level advisors may be assigned as low as brigades.

Coordination

3-33. At the operational level, the advisor may provide guidance to the chief of the local government or senior military commander on the employment of military and paramilitary forces assigned to the AO. The coordination of all military, civilian, and OGA civic action assets assumes increased importance. The realization of IDAD goals depends largely upon the operational-level advisor's capabilities. Organization at this level varies according to the population, economic development, insurgent activity, HN government presence and capabilities, and security posture and concerns.

Training

3-34. HN counterparts may be resistant to training at the operational level, and the SF advisor needs to be persistent in developing sound training programs. HN forces may prefer certain types of training to the exclusion of others. SF advisors must create balanced training programs to address the deficiencies and needs of the HN force (rather than its preferences). SF advisors establish training objectives and maintainable standards, focus on combat effectiveness, and praise training accomplishments.

3-35. SF advisors must emphasize the need for continuous training and encourage counterparts to maximize training opportunities. Counterparts must understand that the training mission needs—
- Military discipline and leadership.
- Marksmanship and small-unit tactics.
- Health, strength, and endurance.
- Proper maintenance.
- Rehearsals.

3-36. SF advisors providing training to paramilitary, police, and other civilian forces must coordinate with other SF and advisory personnel in the AO, and with cooperating U.S. agencies covering PSYOP, agricultural improvement, medical service, PRC, and similar activities.

3-37. SF advisors should train their counterparts to utilize the training resources that they control, to request additional resources from superiors, and to properly allocate resources to subordinate forces. Advisors should stress the importance of such resources as ammunition, films, and training aids, and show how to construct and use field-expedient training devices and facilities. Advisors should encourage them to conduct staff visits and actively supervise unit training. Advisors also should consider—
- Encouraging counterparts to allocate appropriate time and effort to intelligence training—a subject that is frequently ignored by nonintelligence units.
- Establishing training centers to fulfill the requirements for continuous training by rotating all units through the centers in short cycles.
- Employing MTTs when new weapons or tactics are introduced.
- Initiating basic training in areas where the state of unit training is poor.
- Encouraging exchange training between U.S. and HN forces and elements.
- Focusing on the training of unit leaders. It may be necessary to conduct separate officer and NCO schools and classes to better prepare individuals to train and lead their units.
- Assisting the counterpart in establishing training policies and SOPs for units.

- Conducting timely after action reviews (AARs).
- Constructing training villages with caches and booby-traps.
- Developing reaction ranges, close-combat ranges, and infiltration courses to inject realism into training.

3-38. Finally, operational-level advisors must guard against the tendency of counterparts to withdraw units from scheduled training cycles for less important assignments. This disrupts the effectiveness of training efforts and erodes the credibility of training programs.

TACTICAL-LEVEL ADVISOR

3-39. Tactical-level advisors normally advise military units at or below the brigade level and provide counsel and assistance across full spectrum of operations. They advise and assist counterparts in developing unit combat effectiveness and serve as liaisons between HN and U.S. combat, combat support, and combat service support forces. Tactical-level advisors must have a working knowledge of the—

- Tactical air-ground control system.
- Air request nets as integrated with the U.S. Air Force (USAF) and HN air force nets.
- Capabilities, limitations, and operations of the Army, U.S. Navy (USN), USAF, United States Marine Corps (USMC), and their HN military counterparts.
- Organization and procedures pertaining to combined operations.
- Military assistance advisory groups (MAAGs), MILGPs, and MAPs.

INTELLIGENCE

3-40. In order to promote national policy and attain overall objectives, it may be necessary for SF advisors to plan and conduct intelligence training. The SF advisor should assist the counterpart in developing a local intelligence collection program, training intelligence personnel in their respective specialties, and properly utilizing trained intelligence personnel. Advisors may accomplish this aim by—

- Assisting in the establishment of an operations center to coordinate intelligence efforts.
- Maintaining liaison with police and intelligence agencies responsible for countersubversion.
- Providing intelligence support and FP information to U. S. personnel working at other levels.
- Establishing secure and reliable communications channels.
- Preparing daily reports on intelligence-related training and advising.
- Assisting in the development of effective procedures for the collection and dissemination of intelligence information.
- Assisting in the establishment of an adequate security program to safeguard against subversion, espionage, and sabotage.
- Encouraging and assisting the establishment and maintenance of a source control program.
- Assisting in development and achievement of intelligence training for selected, qualified personnel.

3-41. The SF advisor should be familiar with area studies, area assessments, and the special operations debrief and retrieval system (SODARS). Pertinent documents should be compared to detect trends or changes. The advisor also should evaluate the following:

- HN intelligence staff section, its SOPs, and its effectiveness.
- Personalities of counterparts and other persons with whom business is conducted.
- Chain of command and communication channels of the HN unit.
- Intelligence projects initiated by HN predecessors.
- Intelligence projects that SF predecessors believed should have been initiated by the HN.
- Advisor communication channels.
- Reference material available from other intelligence agencies.

3-42. Advisors should prepare and maintain a list of essential elements of information (EEIs) and insurgent indicators, if appropriate. Advisors should determine if—

- Trained subversive insurgent leaders have been discovered.
- Evidence exists of an underground insurgent organization.
- Efforts exist to create or increase civil disturbance and dissension.
- An insurgent psychological campaign is ongoing against existing or proposed government policies and programs.
- Attempts are being made to provoke the government into harsh measures (such as strict PRC).
- Assassinations and kidnappings of local political leaders, doctors, or schoolteachers are taking place.
- Guerrilla actions are occurring.
- An appreciable decline exists in school attendance.

3-43. Advisors may be called upon to provide guidance and assistance in counterintelligence (CI) activities. When acting in such roles, advisors should attempt to answer the following questions:

- Is intelligence information disseminated on a need-to-know basis?
- Are security precautions observed?
- Is access to sensitive areas positively controlled?
- Are cryptographic systems available and used in transmitting classified information?
- Do personnel follow proper communications procedures?
- Are personnel with access to classified information properly cleared? How thorough or effective is the investigation process?
- Are security inspections of installations conducted at regular and irregular intervals?
- Are periodic security lectures conducted?
- Does the counterpart have a covert CI program?
- Does the degree of coverage provide reasonable assurance of gaining knowledge of insurgent intelligence, subversion, or sabotage within the area?
- What means of communication are employed and do they jeopardize the security of the source?
- How much time elapses between a source's acquisition of information and the submission of a report? Does the elapsed time allow for reaction by friendly forces?
- How is the reliability of a source determined? Is reliability (or lack thereof) considered in evaluating information?
- How do counterparts evaluate the accuracy of information received from the source?
- How do counterparts protect operations against—
  - Double agents (agents working for two or more opposing intelligence agencies, only one of whom knows of the dual relationship)?
  - Dual or multiple agents (agents reporting to two or more agencies of the same government, which may result in false confirmation of information)?
  - Confusion agents (agents fabricating information to mislead friendly forces)?

HUMAN RIGHTS AND MISCONDUCT

3-44. SF advisors must stress the consequences of mistreating suspects, prisoners, or other persons taken into custody. These persons must be treated IAW Article 3 of the Third Geneva Convention (GCIII). Article 3 requires care for the sick and wounded, protection of prisoners and detainees of all types from abuse or other harm. Murder, mutilation, and torture are expressly forbidden, as is humiliating or degrading treatment. Sentences and executions may not be carried out unless judgment has been pronounced in the case by a regularly constituted court. SF advisors must never be active participants in the conduct of such punishment.

3-45. U.S. Soldiers must not become involved in atrocities and must strongly discourage all such activities. Furthermore, they must explain to their counterparts that they are obliged to report any atrocities of which they have knowledge.

ROLE SHOCK

3-46. Role shock results from the discrepancy between the roles individuals expect to play and the roles they actually do play. Role shock may also be a product of the tension created by individuals trying to do jobs themselves versus advising others how to do the jobs. Unlike culture shock, which is usually of relatively short duration, role shock tends to increase until about the anticipated mid-point of the tour. Furthermore, role shock seldom disappears completely until redeployment. The symptoms of role shock are similar to those produced by other stressful life events.

3-47. Many SF Soldiers on deployment become entangled in the complexities of bureaucracy—both foreign and American—and become increasingly frustrated when seemingly familiar appearing things fail to respond in expected and predictable ways. This continuing frustration can be a key contributor to role shock.

3-48. SF advisors are not deployed to an AO to exercise technical specialties. Rather, advisors are sent to guide host nationals on how to perform certain tasks. SF personnel may be required to help their counterparts design and implement an organization, a process, or a procedure. Solutions to the attendant organizational political, social, and economic problems require a thorough understanding of many variables outside a Soldier's usual military skill set. In particular, it requires knowledge of and skills in the techniques of advising and facilitating the efforts of others. Whereas it requires certain skills to come into an area and extinguish a fire, a completely different skill set is required to build a fire department and a sustainable fire-prevention program.

3-49. The SF Soldier's role—if executed properly—is rarely "pure." In order to accomplish the mission, the SF advisor must be professionally competent. However, the SF advisor also must be a realist, negotiator, teacher, creative thinker, defender of the long view, and organizer of aid and assistance from other agencies. The advisor needs to represent an objective view by considering problems in a total context and applying experience and sound judgment. The problems facing an advisor are at times made more complex because he must address a range of expectations—from his home unit, from the HN, from his team leader, and from U.S. missions or agencies. At times these expectations may not be fully compatible.

3-50. Although many SF Soldiers undergo some degree of role shock, over time they learn to take the personal, living, and social conditions in stride. More challenging are the problems that arise in connection with their jobs—their professional roles, relationships with colleagues and indigenous peers, personal achievement, self-development, self-determination, and self-image.

3-51. The most frequent complaints voiced by SF Soldiers pertain to the nature of the mission, the nature of the host government, the relationships with counterparts or coworkers, and the lack of self-determination. A significant number of SF Soldiers find that the duties and activities in which they engage during deployments are at least somewhat different from those expected—the actual duties and responsibilities are greater in scope, involve technical work outside of their assigned specialties, and require honed administrative (rather than military) skills.

3-52. Many personnel tend to distinguish between what they consider to be their professional military role and their actual work role. They may exhibit self-confidence as professionals, but have problems in carrying out their work. They perceive their work role simply as those activities carried out during working hours, whereas their professional role encompasses only those activities that require the technical skills associated with professional training. For example, whereas an SF engineer sergeant (MOS 18C) may consider the engineering functions of a project as part of his professional military role, the administrative functions of the project—regardless of their importance—might not be afforded the same professional attention. Because the job of communicating ideas requires a wider range of interpersonal and cultural sensitivity skills than those normally considered "professional," the typical SF advisor finds that his work role behavior differs from that with which he is familiar.

3-53. SF advisors should limit their activities abroad to imparting information and providing advice; however, many assume a performance role. These advisors may rationalize this because actually doing an activity provides immediate and visible results. Advising and supervising is slower, less conspicuous, and more difficult. Therefore, the role challenges become so overwhelming that some advisors retreat into the familiar repetition of more product-oriented activities.

3-54. As with culture shock, an individual that has successfully adapted to HN practices may be vulnerable to a second role shock upon their return to their home station. They may find that they have learned expectations and approaches to working that are quite inappropriate in the United States.

AMBIGUITY OF THE PROFESSIONAL ROLE

3-55. Advisors may differ greatly in their perceptions of structure. Some may regard a rigid structure as guiding, comforting, and facilitating, whereas others prefer the freedom and creativity afforded by more fluid organizations and assignment. Most advisory assignments are ambiguous. What an advisor is expected to accomplish is not always made clear. Furthermore, What an advisor is assigned to achieve may conflict with what he has trained to do, what others expect of him, or what he wants or expects to do.

3-56. Factors unrelated to an individual's tolerance for ambiguity may introduce additional role confusion. These include the administrative context of the assignment, HN attitudes, other team members, and the influence of predecessors and counterparts.

3-57. The United States and the HN may spend months—even years—negotiating a project. Still, the overall objectives or definition of the mission to be accomplished might be expressed in very general terms. Various interpretations of what is to be accomplished arise. Even more views surface regarding the most appropriate means of reaching those goals.

3-58. Many months may elapse between the time the project is agreed upon and the first advisor arrives on the scene. In that time, dramatic changes may occur in objectives and in the staffs of the American mission and the HN. The arriving SF team may find that the local nationals openly or covertly disagree with what they are supposed to accomplish and how they are to go about it. Advisors may perceive that the host government (or individual nationals) entered into assistance agreements only to get economic aid or to achieve some unrelated objectives.

3-59. Advisors can be expected to ponder their role when confronted with such scenarios. More specifically, an advisor may become convinced that the situation lacks the prerequisites for the kind of assistance he can be expected to render. The advisor may resist compromise in the technical approaches or levels of performance that the situation demands. Instead, the advisor may direct his talents and energies into activities that—in his mind—preserve or maintain his professional image.

RELATIONS WITH COUNTERPART

3-60. There exists an almost implicit assumption in most technical assistance projects that the SF advisor will work closely with a host national whom the HN or military organization designates as a counterpart. Although this arrangement appears to be simple, often it is unclear how the role is to be performed in a way that will conform to the expectations of both the American sponsors and the HN.

3-61. Ideally, the counterpart provides a working context for the SF advisor and helps him to understand and adjust to the new culture. Historically, those with good working relationships with their counterparts make better adjustments than those who have no counterparts or bad relations with their counterpart. Good working relationships may be difficult to establish and maintain. Most advisors approach their assignments as teachers and mentors. The counterpart may be resistant to such a relationship, particularly if he was expecting someone to take over and do all or part of the job. In addition, many counterparts want their advisors to share in at least some of the blame if something goes wrong.

3-62. The United States has little control over the availability or capability of counterparts. The assigned individual may be technically incompetent, lack in professional commitment, have other personal or professional interests, or be unavailable when needed. In some cases, numerous counterparts are assigned and rotate in their dealings with the advisor. A single counterpart may not be assigned until the SF advisor has been on the scene for some time. In some scenarios, a specific counterpart is never assigned.

PARTICIPATION IN HOST-NATION BUREAUCRACY

3-63. Although some Americans might disagree, the United States maintains a system that is—by world standards—relatively permissive and nonbureaucratic. As such, advisors may become aggravated when confronted with or obstructed by indigenous bureaucracy, red tape, administrative centralization, and a cumbersome decision-making process in the host government. Cultures that place great emphasis on consensus in the decision-making process can be particularly frustrating for advisors. In order to be successful, the advisor must strive to understand the bureaucracy in which he must operate.

3-64. Communication challenges in overly bureaucratic societies further complicate advisor efficiency. Upward communication in traditionally authoritarian society presents a particular obstacle. Advisors using a foreign language or working through interpreters may fail to identify decision makers, decision-making criteria, and the communication roles held by various levels of military professionals.

3-65. Advisors should recognize the value of being or knowing "the middleman." This role is sometimes one of the key contributions an outsider can make. Middlemen are able to move in, around, and out of organizations somewhat independently of the local hierarchy, protocol, and customs. They can link persons and offices that might otherwise have difficulty in communication. Many SF Soldiers dislike this role and refuse to practice it; however, the middleman can be a significant catalyst in any advisory operation.

RESPONSE TO AMERICAN ADMINISTRATION

3-66. As frustrating as HN bureaucracies may be, SF advisors may become even more irritated with American organizations. Being deployed overseas makes certain individuals hypersensitive to difficulties in organization and administration. Most Soldiers perceive organizations and organization charts as logically conceived structures that operate rationally. They frequently fail to take into account the distances, lead times, diplomatic issues, and other problems involved in establishing and maintaining a functional organization. Moreover, many of the Americans who manage these foreign based bureaucracies are inexperienced in dealing with the military. These issues are further complicated by the usual tensions between HQ and field staffs and between administrators and technical specialists. Expectations, values, and styles of behavior vary from group to group.

3-67. This bureaucratic frustration is not unique to SF advisors. Many local nationals criticize U.S. missions for exercising too much program guidance and control, as well as for its general performance, ineffectiveness in achieving goals, direct interference with the substance of programs, and excessive bureaucracy.

3-68. An SF deployment usually requires more than mere performance of a professional role; it usually implies that the advisor is expected to bring about organizational and individual changes in behavior. This may mean trying to change how nationals conceive and play a particular professional role. It also may require modifying the organizational and political context in which the local national advisor operates. It frequently involves all of these and more, such as the effective employment of the professional role in stimulating or implementing broad institutional or national programs of social, economic, political, or technological development.

3-69. Deploying personnel may not be familiar with the tasks of development and the vast range of problems typical of underdeveloped areas. They come in eager to contribute their knowledge and skill, and anxious to make maximum progress in the limited span of their assignment. In so doing they may fail to recognize that the factors that frustrate them—such as the lack of professional skills and facilities—are actually their reason for being there.

DIVERSITY OF PERSONNEL

3-70. Organizations attract individuals who differ in abilities, knowledge, skills, work ethics, social behaviors, values, and attitudes. Advisors must be able to recognize personality types and develop techniques to cooperate effectively with each type. Figure 3-2 identifies five distinct personality types.

Type	Description
I	Professionally oriented. Most are abroad for the first time.
II	Oriented to interpersonal and social approaches in the work role. Majority have prior overseas experience.
III	Oriented to the administrative process of technical assistance. Most are first-timers with educational administration experience.
IV	Oriented more to the job and the bureaucracy than to the problems, people, or administrative processes of assistance. Most have long service in government or educational institutions.
V	Chiefly concerned with adventure. All first-timers.

Figure 3-2. Personality types

FINAL POINTS TO CONSIDER

3-71. Despite the frustrations and challenges of overseas deployments, most SF advisors reflect upon them as unique and memorable experiences. They count their memories and personal rewards in terms of personal achievement, travel, adventure, excitement, and cross-cultural friendships.

3-72. More perceptive individuals become introspective and admit to such personal dividends as growth in maturity, patience, tolerance, and self-understanding. Most find it much more difficult to assess how much they really achieved while abroad, some are dissatisfied with the amount of progress they were able to make, and a minority feel their talents were not used or were misused. Similarly, some may be disappointed with the degree to which their advisory experience is recognized or rewarded upon return to the United States.

3-73. Even so, most advisors express positive attitudes toward their work and its importance. Despite the problems of cross-cultural work, in most cases advisors experience tremendous personal and professional growth in the course of developing other people, institutions, and countries.

3-74. Overseas deployments make it possible for Soldiers to adjust and learn, to make their contributions, and to return with new attitudes, points of view, knowledge, and skills. This is the important dividend—a dividend expressed in the numbers of people who can not only carry out their professional roles under field conditions but can move effectively in and out of intersocietal relationships.

Chapter 4

Cross-Cultural Negotiations

Because negotiation is fundamental to problem solving, the study, development, and honing of negotiation skills are critical to the SF advisor. The principal form of negotiation that the SF advisor is likely to conduct is cross-cultural negotiation. Cross-cultural negotiation brings with it a series of unique challenges and techniques. Culture fundamentally affects language and behavior. It also significantly impacts the way people handle conflict. When it comes to negotiation, one culture may prefer to use a competitive style (win–lose), whereas another culture may prefer compromise or accommodation (win–win). Cultural differences may lead to a conflict between what the parties expect and what their families and communities expect. This has an obvious impact on negotiating behavior.

Negotiations among members of the same culture can be stressful; negotiating with members of other cultures can be exceedingly difficult. Logically, working with other cultures is a basic skill for the SF Soldier and an absolute requirement when acting in an advisory or foreign liaison role. Understanding the factors discussed will help minimize the difficulty of cross-cultural negotiation.

PERCEPTIONS

4-1. Most Americans have certain preconceived ideas about people from other cultures. These perceptions may not be factually based, but they do exist nevertheless, and they influence the way people approach negotiations with foreigners.

4-2. Similarly, foreign negotiators have certain perceptions about American negotiators. Again, these perceptions may not be based in fact, but they too exist. Advisors need to know how other cultures perceive American negotiators so that they can adjust their negotiating style accordingly. The successful advisor must find ways to capitalize on the positive preconceptions foreign negotiators have of Americans and find ways to neutralize the negative preconceptions they may have.

4-3. Research indicates that different cultures hold different perceptions of Americans. For instance, most cultures think that Americans are hard working. The Japanese, however, don't associate this trait with Americans at all. Most Japanese perceive Americans as rude, and many foreigners think that Americans are culturally insensitive. They believe that most Americans are interested only in their own culture, language, methods, and customs. This perception is based almost entirely on two key factors. First, most Americans speak only one language—English. Many U.S. negotiators believe it unnecessary to learn any foreign language; if negotiators from other cultures want to do business, many Americans expect them to do so in English. Second, most Americans have distinct business customs from which they are unwilling to deviate. These standards and customs are forced upon foreign negotiators in practically every business transaction.

4-4. Because these perceptions can create negative attitudes, U.S. negotiators must be sensitive to the preconceptions of foreign negotiators and work to counter any negativity. This is not to imply that negotiators must become fluent in every foreign language; by learning a few simple phrases, U.S. negotiators can show that they are at least somewhat interested in and respectful of the local culture and customs. At the very least, negotiators should learn the following phrases (or their equivalent) in the local language(s):

- Hello.
- Goodbye.

- Yes.
- No.
- Please.
- Thank you.
- Good morning.
- Good evening.
- I hope to see you soon.

4-5. In addition to common terms and phrases, negotiators must learn the proper way to address people. In recent years it has become common for strangers in the United States to address each other by their first names. This custom has not yet taken hold in other cultures—most prefer to be addressed by some honorific title. Many counterparts consider it rude if negotiators address them by their first name. This is especially true of those who hold important government positions, academic titles, or military ranks. In some African societies, for example, it is common to address government officials as "chief," much as U.S. officials are often addressed as "the honorable Ms. Smith." Negotiators must research the local-language equivalent of military ranks and common titles (for example, Mr., Mrs., colonel, professor, and doctor). Formal titles should be used until the counterparts invite negotiators to use first names or other titles.

4-6. Negotiators may further counteract negative perceptions about American cultural sensitivity by having information about them, their organization, and their mission translated into the foreign language. There are numerous software packages that can translate important messages into French, Spanish, German, Italian, and several other languages. It is not necessary to translate every document into the foreign language; often a translation of one or two key documents will suffice. This small gesture, requiring minimal effort, may be enough to show foreign negotiators that the U.S. negotiators appreciate their local language and customs.

INDIVIDUALISM

4-7. The United States encourages a largely individualistic culture. As such, the typical American negotiator prefers to enter into negotiations alone or, if required, as part of a very small negotiating team (two or three people, at most). This behavior is largely a byproduct of U.S. culture, which focuses on the individual, performance, initiative, and accomplishments.

4-8. Most other cultures of the world place less emphasis on the individual and more on the group. Cultures that are extremely group-conscious include the Latin American countries of Mexico, Argentina, Brazil, Venezuela, and Colombia. Cultures in the Pacific Rim, such as Japan, Malaysia, Hong Kong, China, and Taiwan are also group-oriented. Most societies in Southern Africa also emphasize group decision-making. In cultures where the group—not the individual—is responsible for decision-making, negotiations can drag out for long periods of time. Figure 4-1, page 4-3, provides a brief snapshot of the cultural views of the individual versus the group.

4-9. When dealing with group-oriented cultures, U.S. negotiators may wish to reconsider the inclination to enter into negotiations alone (or with very small group). In some scenarios, the arrival of a single negotiator indicates to counterparts that the individual (and, by default, the United States) is ill-prepared and unprofessional. In their view, a single negotiator does not have the required experts to provide advice or support during the negotiation process. In some cultures—Russia, for example—sending a single negotiator or a very small team may be taken to mean that the other side is not serious.

4-10. Additional challenges face unaccompanied negotiators. When an individual attempts to negotiate on their own with groups of ten or more, they may become flustered. Rather than one-on-one dialogue, the sole negotiator must absorb pressure from many persons. Sole negotiators must convince every member of the group. The individual negotiator must divide his focus among the team members of the other side, whereas they can focus on him alone. This can prove to be a nerve-wracking experience.

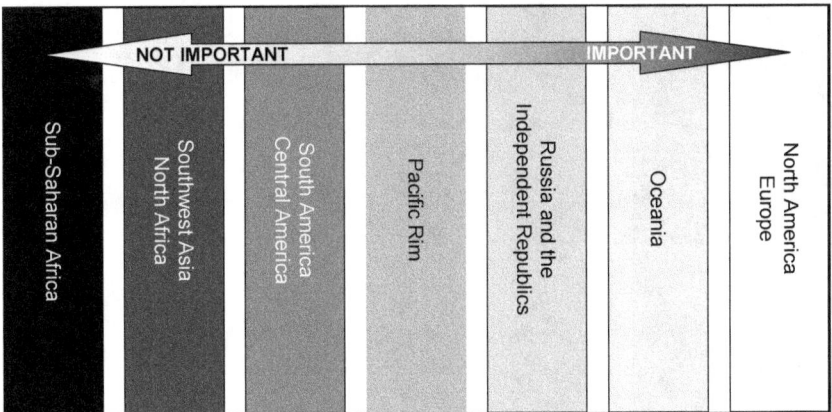

Figure 4-1. Individuality

4-11. Negotiators should increase the size of their team when conducting business with such cultures. If possible, negotiators should find out in advance how many people their counterparts plan to include in their team. The number of U.S. team members should be increased to mirror the counterpart, and this number should include subject-matter experts (SMEs) to assist during negotiations.

4-12. Unlike the group-think mentality of many cultures, delegates from Europe, Canada, and (Caucasians in) Southern Africa emphasize individual decision-making. Still, the decision-making processes employed in these regions are not quite as individualistic as that of U.S. negotiators. When negotiating with people from these cultures, U.S. negotiators may need to adjust their approach slightly. Although they should concentrate on the chief decision-maker, they should not entirely overlook the other members in the team. These individuals will still influence the decision-maker during private discussions and caucuses.

PUNCTUALITY

4-13. Americans place a great emphasis on punctuality, viewing it as an indication of an individual's basic ability and commitment. Professionals from Australia, the Benelux countries, France, Germany, Switzerland, Sweden, the United Kingdom (U.K.), and Japan emphasize punctuality even more than Americans. Conversely, time is considered relative for those in Mexico and other Latin American countries. Members of the most African cultures cannot understand why others have to conduct all their activities according to a clock. Appointments often start late, are rescheduled, or are canceled altogether. Figure 4-2, page 4-4, provides a snapshot of the cultural views of punctuality.

PACE OF NEGOTIATIONS

4-14. The pace of negotiations in the United States is faster than in most other cultures; consequently, the negotiating process between Americans is usually much shorter. In the opening phase of negotiations, American negotiators normally do not spend much time on introductions and getting to know their counterparts. Little attention is given to building rapport or creating a positive and relaxed negotiating climate. American negotiators typically do not show much interest in learning about the roles of the team members in the counterpart's negotiating team. In fact, an American negotiator may not even spend much time introducing himself to the other members of his own team. American negotiators tend to attack the task of negotiating as quickly as possible.

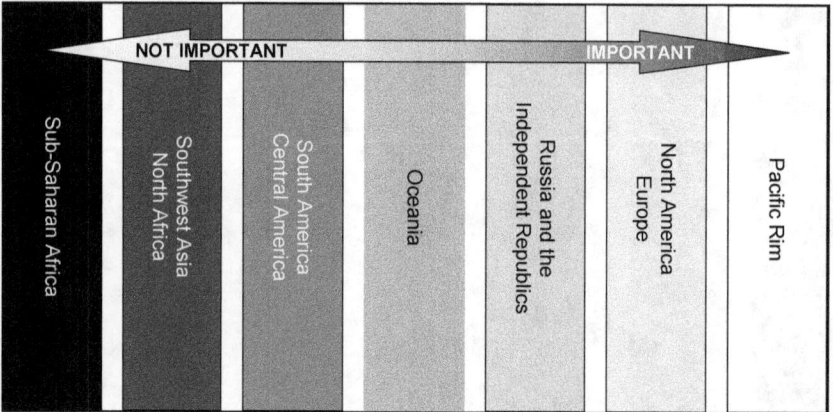

Figure 4-2. Punctuality

4-15. Although the results-oriented approach is effective in many U.S. scenarios, it may create the impression with foreign counterparts that the U.S. negotiator is untrustworthy—that he is trying to force a decision at the expense of the other side. Foreign counterparts may feel that the U.S. negotiator is trying to exploit both them and the situation. Such negative perceptions may lead the foreign negotiator to avoid making any deals. The foreign negotiator—unlike his American counterpart—often does not want to simply arrive at a settlement or conclusion; he wants to build a positive relationship for future negotiations.

4-16. Negotiators from many other cultures spend more time on relationship issues and building rapport, particularly at the beginning. If U.S. negotiators do not consciously force themselves to slow this important phase of negotiations, they will fail to get vital information from their counterparts, create distrust, and weaken their chances of gaining vital concessions. Throughout cross-cultural negotiations, American negotiators resist temptations to make early concessions in order to move the negotiation forward or to demonstrate willingness to compromise. American negotiators who rush may make unnecessary concessions and actually weaken their position. Interestingly, research suggests that the party who makes the first concession usually gets the worst part of a deal.

4-17. Negotiators in the United States tend to close negotiations much faster than negotiators in other cultures. Often working under rigid timelines, U.S. negotiators must produce results quickly so that they can turn their attention to other tasks. The closing of a negotiation is looked upon by many other cultures as the time to cement the relationship. An American negotiator who rushes away from a successful meeting may leave the foreign negotiator with a less-than-desirable final impression.

Note: A counterpart may offer to help with travel arrangements as an act of courtesy. Negotiators should use caution in accepting this offer. Counterparts who make the travel arrangements become privy to how much time the U.S. representative has to negotiate. The foreign negotiator may use this information to add additional pressure as the scheduled departure time draws near.

4-18. Negotiators from Canada, the Benelux countries, Sweden, Switzerland, and the U.K. maintain a pace similar to that of Americans. The negotiating pace in countries such as Italy, Spain, and even Australia is somewhat slower than in the United States. French, German, Russian, and Japanese negotiators are used to a much slower negotiating pace. Negotiators from the Pacific Rim, Latin America, and most of Africa proceed the slowest. American negotiators must significantly slow their pace when negotiating with persons from these cultures. Figure 4-3, page 4-5, provides a snapshot of cultural views of negotiation pace.

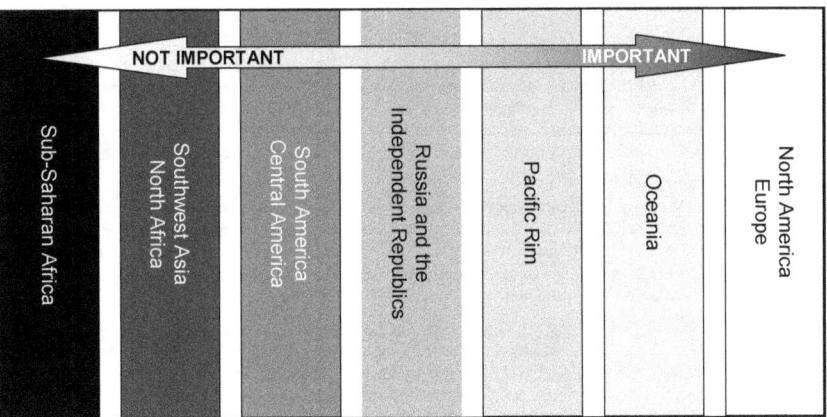

Figure 4-3. Pace

RELATIONSHIP BUILDING

4-19. Americans are very competitive during negotiations and are inclined to stress short-term results. Interpersonal relationships are of minimal importance to American negotiators. Building long-term relationships only occurs after the successful completion of negotiations.

4-20. Countries in Western Europe vary somewhat in the importance they place on establishing business relationships. German and French negotiators are very similar to Americans, placing little value on the development of long-term relationships. British, Scandinavian, and Swiss negotiators, however, display a slightly larger need to build relationships. Spanish and Italian negotiators have an even greater need to establish good relationships with their negotiating counterparts.

4-21. Except for Southern Africa, negotiators from the African cultures often stress the need to build relationships among themselves, but do not display a similar emphasis with negotiators from non-African cultures. When conducting negotiations, most Eastern European cultures do not value personal relationships as much as Americans. This is unlike the neighboring Russian cultural region, where negotiators display a slightly greater need for personal relationships—about as much as that of their American counterparts.

4-22. In Latin America, the Middle East, and the Pacific Rim, personal relationships rate high among the needs of negotiators. Friendship opens the door to a successful negotiation. Negotiators often expect to get together before negotiations so that they can get to know one another on a social level. Even during negotiation, members of these cultures spend a long time on general conversation before they introduce business. Negotiators from these three regions spend a great deal of time on the opening phase of negotiations. They first need to know the person with whom they are negotiating, and they must have a high level of to trust in that person before they start making deals.

4-23. When dealing with persons from Latin America, the Middle East, and the Pacific Rim, American negotiators should plan first to engage in small-talk. Subjects such as politics, race, religion, and gender issues should be avoided. These topics seldom help to build relationships between strangers. Instead, American negotiators should try discussing the foreign country's history, cultural heritage, traditions, beautiful countryside, contribution to the arts, economic successes, and popular sports. Questions about local restaurants are usually safe, neutral topics to begin a conversation. Negotiators must be prepared to discuss typical American traditions, sports, and cultural heritage; however, they must be careful not to go overboard with talk about America. It may come across as pompous and overbearing.

LANGUAGE

4-24. Business, government, and military personnel in Western European countries—particularly Germany, France, and the Benelux countries—commonly speak English. In France, however, even if the counterpart is fluent in English, negotiators are expected either to speak French or to use an interpreter. Most French do not like to speak English; they are very proud of their language. If a French negotiator does agree to use English, he will make it known that he is doing his American counterpart an enormous favor.

4-25. In some Western European nations—Spain, Portugal, and Italy, for example—most business, government, and military people do not speak English. Unless fluent in the local language, negotiators will need an interpreter. Counterparts in these countries also will expect American negotiators to present them with detailed written proposals in the local language. Interpreters also are required in any countries or regions where English is not a common language, such as Russia, Eastern Europe, and China.

4-26. As with most cultures, most Middle Eastern and Central Asian cultures appreciate foreigners who take the trouble to learn a few simple courtesy words (such as hello, goodbye, and please). However, most Arabs greatly treasure spoken Arabic. They often prefer that foreigners who are not fluent refrain from using more than a few, basic courtesy words. Most Arab businesspeople, government officials, and military officers speak at least some English; many are quite fluent.

4-27. English is widely spoken in business circles of Latin America and the Pacific Rim countries, but not necessarily within local government and military circles. English is a common language in Sub-Saharan Africa because of the vast number of tribal dialects. English and French are common among the educated in sub-Saharan Africa, with English being more common in East Africa. English is an official language of Kenya. The western portion of sub-Saharan Africa has more French speakers, who often consider speaking French the mark of an educated person.

CONDUCTING CROSS-CULTURAL NEGOTIATIONS

4-28. The basic elements common to all negotiations must be applied in way that allows for the differing cultures of the participants. Elements that must be considered include the following:

- Opening strategies.
- Directness.
- Strengthening behaviors.
- Movement.
- Power.
- Face-saving.
- Formal agreements.
- Mediation.

OPENING STRATEGIES

4-29. The opening strategies of negotiators differ from one culture to the next. In the United States, negotiators tend to open with offers or demands that are very different from their final positions or expectations. They leave a healthy margin to bargain. Many other countries use a similar approach to negotiations; however, the application of this strategy may be very different.

4-30. Negotiators from most Pacific Rim countries also tend to open high, but not as high as American negotiators. Negotiators from these countries might not allow themselves as much bargaining room as American negotiators might expect.

4-31. Negotiators from the Middle East and Russia usually open with high to extremely high demands. Russians are well-known for opening with extreme demands or offers; some may even strain or exceed credibility. Therefore, they allow themselves a tremendous amount of bargaining room. American advisors must be cognizant of this practice so that they can build enough "fat" into their own negotiating range.

4-32. Doctor Henry Kissinger adapted his negotiating style to match his Russian counterparts. In his negotiations with the U.S.S.R., Dr. Kissinger always opened with extreme demands and offers. If agreement must be reached somewhere between the two side's opening positions, he reasoned, it would make little sense to open with a moderate or reasonable demand. With the distance between opening positions broadened, Dr. Kissinger allowed himself sufficient bargaining room.

4-33. African negotiators tend to open with high to extreme demands. This is especially evident in labor negotiations. In other regions, such as Western Europe (less Spain and Scandinavia) and Latin America (less Mexico), negotiators open with moderate demands or offers that are close to their walk-away positions. American negotiators may expect them to move slowly and to make small concessions.

4-34. Regardless of the cultural differences between Americans and negotiators from other countries, it generally is wise to aim high. High opening positions lower the counterpart's expectations. They convey a silent message that the American negotiator believes in his case. They also leave enough bargaining space to allow the counterpart to win concessions without materially affecting the final outcome. This allows the foreign negotiator to save face and report back to his superior that he was successful.

DIRECTNESS

4-35. American negotiators are extremely direct. They ask counterparts direct questions, such as: "How do you feel about my proposal?" American negotiators are time-driven and want fast results. Because this approach is almost entirely unique to American and certain European cultural regions, it can create tension and difficulties. Negotiators from most other cultures do not appreciate such directness. They perceive Americans as pushy and they resent direct behavior.

4-36. Negotiators from Latin America, Russia, Africa, and the Pacific Rim are less forthright than Americans. In particular, members of the Pacific Rim cultural region—particularly China and Japan— negotiate in a very indirect manner. The true message from a Japanese negotiator, for example, is usually found between the lines. It must be extracted from the general context of the negotiations. A Japanese negotiator may agree with an American negotiator without ever really saying yes. Similarly, the same individual may say yes without really agreeing; he may simply be indicating that he understands the message.

STRENGTHENING BEHAVIORS

4-37. American negotiators prefer to deal with one issue at a time. In the United States formal presentations are favored, backup information is accurate, differences are dealt with directly, and detailed discussions of issues are common. Being results-driven, U.S. negotiators try to get through the negotiations as quickly and efficiently as possible.

4-38. The negotiating practices of Europeans are somewhat similar to the United States. They candidly express disagreements; however, they do so politely. Europeans tend to be more precise with facts than their American counterparts. They expect greater detail in presentations and will take time to analyze the data very closely. They appreciate conceptually strong presentations. They are argumentative and like to debate issues to search for flaws in the logic of the opposing position. If they discover such a flaw, they will focus on it and fully exploit it. Any hesitation in answering their questions is normally interpreted as a sign of uncertainty, weak preparation, unprofessional behavior, or, even worse, as an indication of deceit.

4-39. African negotiators are less interested in the underlying logic of a position and are more prone to focus on specific facts and the details of propositions. With extensive questioning and debate, negotiations are often long, slow, and frustrating. The regional tendency for collective decision-making helps to slow the process even more. Negotiators often want to consult with persons who may be affected by the negotiations, particularly those who were not present. Although they prefer to discuss groups of issues, they are quick to pick out the good concessions and continue negotiating the ones they dislike. They will usually state any disagreements quickly. At times they may disagree so fiercely that it may come across as rudeness. However, care should be taken in interpreting their gesture. Normally, it is less a negotiating tactic than a normal means of expression. There is a good chance that no offense is intended.

4-40. Russian and Eastern European negotiators also expect rational presentations and will ardently argue the reliability of the facts presented. They often link issues and discuss them in groups instead of addressing issues one at a time. Consequently the American negotiator must be thoroughly prepared and ready to cover multiple issues at a time.

4-41. Latin American and Middle East negotiators are notably passionate and argumentative. Emotions play a significant role during these negotiations. People from these regions express themselves strongly and vividly. They frequently wave their arms, speak very loudly, shake their heads, and throw down their pens to show their astonishment at the opposing side's positions. A negotiator who does not expect this behavior will feel uncomfortable, embarrassed, and perhaps even ashamed of his proposals. American negotiators must be prepared to deal with these emotional displays and not allow them to lead to unnecessary concessions. The best way to accomplish this is to allow the opponent to carry on without reacting to it; it should never be taken personally. The foreign counterpart (probably) does not intend to embarrass or make the American negotiator uncomfortable. It is simply part of his culture. The worst thing an American negotiator can do when encountering this behavior is to begin making concessions. Such actions simply reward the counterpart for his behavior.

4-42. Negotiators from the Pacific Rim prefer to have a large amount of information to help them decide. They use considerable technical detail to back up their proposals, and they expect the same of their opponents. Their negotiating style is reserved; they will quietly and politely disagree. Negotiations can be lengthy. Pacific Rim negotiators—particularly those from Japan, China, and Malaysia—like to carefully analyze data. Also, the group decision-making practiced in most Pacific Rim countries slows the negotiating process because all the members must agree.

Note: One notable exception among the Pacific Rim countries is Singapore, where it is customary to negotiate briskly.

MOVEMENT

4-43. Americans are tough negotiators. They concede points reluctantly and save concessions until late in the negotiations. When they do concede an issue, that concession often is the only one they are prepared to make on the issue. Because they are only prone to make a single concession, they tend to hold out for a long time without budging. When they eventually do move, they normally make the entire concession in a single move.

4-44. Other cultures may have different concession behaviors. The successful negotiator needs to become familiar with these behaviors. If the American negotiator knows what to expect, he can better prepare to adapt the strategy accordingly.

4-45. As noted earlier, consensus is important to African negotiators and they tend to base their decisions on this group consent. This significantly slows the negotiating process and leads to an escalating pattern of concession-making, with the larger concessions made toward the end. Negotiators from the Pacific Rim also move too, group consensus is critical.

4-46. Negotiators from Russia and Eastern Europe tend to take an even harder line than Americans or Europeans. They do, however, move very slowly. This includes the granting of any concessions. Historically, Russian negotiators possess very limited authority and must regularly report back to their principals. Several negotiating sessions, with lengthy periods between, may be necessary to complete an agreement.

POWER

4-47. In the United States, military, business, and government organizational power tends to spread from the top downwards. The most important job in any organization is the most senior officer who has final decision-making power. However, it is common for Americans to delegate much of this power to subordinates of middle rank. Officers lower down the line often are involved in key decisions. A middle-ranking officer may enjoy considerable power in deciding everyday issues. He may also have full authority to negotiate on certain issues.

4-48. In Western Europe organizations, the involvement of middle managers in key decisions is low. Delegation of power is limited. In Eastern Europe, power is even more centralized and bureaucratic. This slows the negotiating process. When conducting business in Europe, American negotiators must ensure that the counterpart has the authority to make decisions.

4-49. In Latin America, the senior officer makes decisions. Middle-ranking negotiators take their cues from the senior member of the negotiating team. Negotiators operating in Latin America must focus on convincing the negotiation leaders of the merit of their proposals.

4-50. Along the Pacific Rim, organizational power tends to be more-or-less evenly distributed among the various levels of management. Decision-making is based on group consensus. The leader of the foreign delegation may not have full authority to make commitments with the American negotiator. Often, however, this individual does have the authority to cancel negotiations. In other words, although he may not have the authority to say yes, he may have the authority to say no.

4-51. In Africa, organizational power often follows the hierarchy of the organization. The most senior manager makes decisions, but he will consult with managers lower in rank. The senior manager will make decisions that he believes are correct, even if it runs counter to the opinion of lower managers. American negotiators should concentrate their efforts on the most senior member while convincing the entire group.

FACE-SAVING

4-52. Although nobody likes to be embarrassed, American negotiators have a comparatively low need to save face during negotiations. Negotiations are won or lost, and the American negotiator moves on to the next project. Negotiators from cultures such as Latin America, Japan, and other Pacific Rim countries show a far greater need to save face. In these countries, it can be a major disaster to undermine the respect and value of your opponent in the eyes of his colleagues. For instance, an American negotiator should never address the person on the other side who speaks the best English. The more senior person in the other team may take this as a great insult—one that is not easily forgiven. American negotiators should never use curses, vulgar expressions, or other expletives during negotiations. Criticism should not be made unless absolutely necessary and should only be done in private. Most importantly, American negotiators must be prepared to make concessions that the opponent can take away as a win or gain.

FORMAL AGREEMENTS

4-53. Americans are known for their willingness to approach the courts for legal assistance. The impact of this phenomenon on negotiations is that foreigners may not trust Americans who want to create extensively detailed agreements. They may see it as the American negotiator's first step toward taking them to court. Because the foreign negotiator wants to build a relationship with the person with whom he is negotiating, he may feel that excessive paperwork is obstructive to building trust. Consequently, an American negotiator must be sensitive about these cultural differences, and should attempt to balance the counterpart's distaste for detailed contracts with the need to secure American interests.

4-54. Negotiators from Latin American, Pacific Rim, and Middle Eastern countries tend to avoid extensive written contracts. They want to create relationships, not paperwork. Trust is the cornerstone of their negotiations. If they do not trust someone, that person will have a serious problem in trying to make a deal with them.

4-55. The American fondness for detailed contracts is shared by other cultures. Negotiators from Europe, for example, also tend to draft thorough contracts. This is also true of many African negotiators. American negotiators should experience few problems when presenting extensive written agreements to negotiators people from these countries.

MEDIATION

4-56. Negotiation in the United States often is viewed as two sides, each attempting to argue one position against the other. The SF advisor taking part in foreign negotiations may be required to assume a unique role—a role that does represent one side against the other. The SF advisor may need to assume the role of mediator between two opposing parties.

4-57. Mediation is a peaceful method for resolving differences and disputes with the help of an outside intermediary. Many cultures fail to appreciate the value of mediation. Because often it is not explained properly, many people do not understand the process. Others mistake it for arbitration, where someone makes a decision for the parties. Culture affects the way people view mediation.

4-58. Cultural differences cause inconsistencies in the expectations of mediators. In Western society, neutrality of the mediator is important. In African society, mediators are expected to provide advice or to offer solutions. If a mediator does not offer advice, the African parties may feel that the mediator is ineffective, and the Western party may feel that the mediator is doing a great job. Similarly, if the mediator does offer advice, the African may feel that the mediator is doing a great job, and the Western person may feel that the mediator is biased.

4-59. If serving as a mediator, the SF advisor must present a neutral position. Presenting neutrality can be difficult when the mediator comes from a different background than the disputants. The two disputants may view the mediator and one another with such skepticism that mediation may not be possible at all. The mediator will have a major task convincing the disputants of neutrality. The mediators must find ways to prove neutrality.

4-60. Cultural differences may affect communication during mediation. For example, eye contact during mediation may be appropriate between disputants of the same culture, but it may be inappropriate between disputants from other cultures. In some cultures, maintaining eye contact is a sign of respect; in others it may be viewed as offensive. Disputants from different cultural backgrounds must be sensitive to the significance of every gesture.

4-61. Mediators must try to guide parties through rational problem-solving stages. This may clash with one (or both) of the parties' cultural decision-making or conflict-resolution patterns. Certain groups use circular reasoning or passionate discussions.

4-62. In cross-cultural conflicts, an enormous imbalance of power may exist, particularly between majority and minority groups. The more powerful party may exert greater influence because of better negotiation skills or greater resources. In order to succeed at mediation, a mediator may try to redistribute the power. When this happens, the more powerful group may conclude that the mediator is no longer neutral and may, as a result, withdraw from the process.

4-63. Mediators must understand that cultural values and biases influence everyone. They must realize that cultural conditioning may be the cause of the parties' negative feelings toward one another. Although the problem-solving approach makes sense to many people, others will revert to a confrontational stance in certain situations. To be successful, mediators must develop the ability not only to see the conflict clearly from his perspective, but also from the perspective of each of the disputing parties.

Chapter 5

Joint and Interagency Environments

The SF advisor will almost always, to some degree, operate in a joint and interagency environment, working alongside other entities of the USG. This environment may include formal military command and control (C2) structures, such as joint task forces (JTFs), frequently will involve working alongside various interagency partners, and almost always will fall under the responsibility of one of the elements of the Country Team of a U.S. Embassy.

CULTURE AND THE OPERATIONAL ENVIRONMENT

5-1. SF Soldiers often are assigned duties that enable or facilitate interdependent operations through interface with joint or Service elements. These Soldiers may serve as staff augmentation or as part of a special operations command and control element (SOCCE). Even in these cases, SF Soldiers are serving as advisors to Service component or joint component commanders and staffs, each with its own culture. Therefore, SF Soldiers should approach their counterparts using the fundamentals of cross-cultural communication. The degree to which SF Soldiers are able to influence the actions of the HQ to which they are attached will be based—to a large extent—on a process of negotiation. The principles and techniques of cross-cultural communications and negotiation also pertain to joint environments. SF Soldiers acting in such capacities should study and analyze the organization with which they are working and attempt to determine the culture(s) of that organization. This holds particularly true in joint and combined organizations.

5-2. In this context, the culture(s) of the supported HQ will either be a joint or Service component culture. Subcultures exist even within each Service components. Examples include Army branches, Naval qualifications (such as air, surface warfare, and submarine warfare), or Air Force tactical, strategic, missile, and airlift organizations.

5-3. A thorough analysis of every component subculture is beyond the scope of this publication. Review of joint publications (JPs), Service doctrine and training literature, and other available source material is fundamental to gaining a full appreciation of joint and Service cultures and subcultures. Carl H. Builder's book, The Masks of War, is one such source.

5-4. The following sections provide brief, broad generalizations of joint and Service culture. They must be adapted to individuals and specific units. The key to effectively operating in the joint arena is to consider joint and Service component elements and personnel as counterparts. This demands an awareness that they come from distinct cultures, knowledge of the particular culture represented, and the application of cross-cultural and negotiation skills. As always, the key to success for the SF advisor begins with the first SOF imperative—understand the operational environment.

JOINT CULTURE

5-5. The information provided on joint and Service cultures is anecdotal. It is derived from numerous interviews with former SF advisors and illustrates a perception of joint culture from an SF advisor's perspective. However, these perceptions represent a consistent theme among those interviewed and are worthy of consideration.

5-6. For a variety of reasons, there is not yet a true, distinct, universal joint culture that can be easily and succinctly defined. There is, however, a great variety of joint doctrine, procedures, and experience in the

U.S. Armed Forces, particularly in the last two decades. Since the terrorist attacks of 11 September 2001, the U.S. Armed Forces have greater experience and comfort in the joint environment. On the positive side, representatives from each Service are generally considered to be experts in their respective fields. One negative effect is that the individual members of a joint HQ may not understand all the terminology and procedures of both the joint environment and the other individual Services, often leading to frustration and stress. Interestingly, as members of a joint command (United States Special Operations Command [USSOCOM]), SOF tend to adapt much more quickly to the challenges of joint environments than conventional Service members.

5-7. Although it is difficult to define a truly joint (or "purple") culture that is universal to all joint organizations, it is useful for the SF advisor to think of joint culture as an amalgamation of Service component cultures functioning at the joint level. The relative importance of each Service culture within this mix varies widely over time—both between and within individual HQ. Relative dominance by one or another service culture within a given HQ follows along functional or organizational lines and varies according to the mission and environment. Such dominance may even vary between phases or aspects of operations. SF Soldiers enabling interdependent operations must observe and analyze joint HQ to determine which cultures are dominant under what circumstances. This analysis becomes the basis for cross-cultural communications with—and the influencing of—participating joint organizations.

5-8. Relative dominance by a component Service culture results from several factors. Careful analysis of these factors, while observing and analyzing the actions and responses of the "target" HQ, can assist the SF Soldier in determining the key aspects of the dominant Service culture. These factors include the—

- Background and Service affiliation of the unit commander.
- Background and Service affiliation of key staff personnel.
- Type of unit.
- Nature of the mission or task.
- Origin and type of subordinate operational units.
- Origin and type of subordinate support units.
- Service expertise associated with a function or aspect of the mission.
- Applicable joint or Service doctrine.

5-9. Each Service has its own distinct culture. Given the importance of the individual Service cultures, a brief review of some of the most relevant points is listed below.

SERVICE COMPONENT CULTURE

5-10. Each Service has its own culture and subcultures. A detailed knowledge of the characteristics of each is the product of research, study, and experience. Service documents (particularly doctrine) are an excellent start point for research, and an advisor should reflect upon his experience and the experience of other advisors in dealing with the different Service cultures.

ARMY

5-11. There are a number of distinct qualities that make up U.S. Army culture. These qualities include adherence to standards, selfless service, a human focus, the view of warfare as an art, and an acceptance of joint and interdependent operations.

Adherence to Standards

5-12. The Army encourages the enforcement of standards and discipline, which are clearly defined in Army regulations and procedures. Although highly disciplined, the average SF advisor may put little emphasis on uniform and appearance, particularly in a combat environment. This would be a mistake, however, when the SF advisor deals with conventional units of the U.S. Army. In such scenarios, the SF advisor would benefit by conforming as much as possible to standard Army policies (for example, wearing an issued soft cap instead of a non-issued baseball cap). Such inconsistencies in appearance may cause tension between conventional and unconventional forces, a tension that is easily avoidable.

Selfless Service

5-13. The Army feels deeply its attachment to the people of the country. This may be attributed to the relative importance of the individual Soldier drawn from the citizenry into the Service. This reliance tends to lead the Army to view itself as a humble servant of the nation. As a result, the Army culture does not promote perceived grandstanding or show-offs, and rejects perceived arrogant behavior. If SF advisors maintain standards of behavior befitting quiet professionals, they should have no problems.

Human Focus

4-64. Often expressed as "equipping the man, not manning the equipment," this focus contrasts with the systems-focused approach of the Navy and Air Force. The Marine Corps shares in the Army's human focus.

View of Warfare as an Art

5-14. Although the Army still applies the tools of Military Science, such as "battle calculus," the Army generally remains adverse to reliance on mathematical models and other purely quantitative tools to predict outcomes or evaluate combat capability. Within certain branches of the Army, such as the Field Artillery, Engineer, and Armor branches, there tends to be more of a focus on the scientific approach. Overall, however, the Army sees warfare more as an art more than as a science, and frequently uses the term "operational art" to describe its approach to operations. This view flows naturally from its human focus.

Acceptance of Joint and Interdependent Operations

5-15. The Army has long accepted the necessity of joint operations. Reliance on the other Services for deployment, logistics, and other support are contributors to this attitude.

NAVY

5-16. It is important for the SF advisor to recognize and understand a number of distinct qualities that make up USN culture. These qualities include an adherence to tradition, an independence of command, and a strong focus on ranks and specialties.

Adherence to Tradition

5-17. The Navy is perhaps the most traditional of the Services. Although this dedication to tradition has certain positive implications, it also leads some conventional Navy personnel to be very resistant to change. SF Soldiers must be respectful of Navy traditions. Likewise, every attempt should be made to articulate ideas and operational concepts in a professional manner, consistent with existing Navy thinking.

Independence of Command

5-18. For obvious reasons, there can be only one captain of a ship at sea. Because of the historical isolation of ships at sea, the Navy tends to vest more discretionary authority in the captains of its vessels than other Services grant to their subordinate commanders. Navy units and personnel tend to be less accustomed to routine external interference and outside presence than other Services. SF Soldiers should be sensitive to this fact, stress their role as an enabler, and avoid the appearance of interfering with the prerogatives of the commander.

Emphasis on Ranks and Specialties

5-19. The Navy is scrupulously conscious of ranks and specialties and maintains distinctions between those specialties. Rank separation is significantly more pronounced than in the other Services. SF Soldiers, accustomed to relative informality, should take care not to inadvertently give a false impression of irreverence for military structure and deference to rank.

AIR FORCE

5-20. There are a number of distinct qualities that make up USAF culture. These qualities include a focus on technology and science, an emphasis on pilots, an emphasis on targeting, and a powerful self-image.

Focus on Technology and Science

5-21. The Air Force is technology-oriented and tends to view warfare as a science that is dominated by interface of humans and systems. Applications of technology are viewed as solutions to most (if not all) problems associated with warfare and national security. The Air Force is generally more inclined to depend on the type and capability of a specific system rather than on their numbers. This reliance on high-cost, low-density equipment may have implications for risk acceptance.

Emphasis on Pilots

5-22. Air Force combat doctrine traditionally is based around the employment of piloted aircraft. As a result, pilots are far more likely than their nonpilot counterparts to have ultimate decision-making authority and a greater understanding of the application of force. Large portions of the Air Force exist to support the pilots in their mission. As such, pilots tend to be more individualistic and tend to have greater social standing. Similarly, other members of the Air Force on flying duty are considered to be more elite than those in support (nonflying) roles.

Emphasis on Targeting

5-23. The Air Force is extremely target-oriented and conducts tactical targeting on an operational scale very effectively and more effectively than any other Service. The technological orientation of the Air Force lends itself to complex systems that match appropriate platforms to targets. The human dimensions of warfare—that is, those that are not related to human-machine interface—can be subordinated to quantifiable target destruction. Speed and precision are valued far more highly than persistence and endurance.

Powerful Self-Image

5-24. The Air Force sees itself as the new combat arm of decision. The Air Force views air power—particularly strategic airpower—as decisive and potentially independent of the other Services. The Air Force views rapid decisive operations as a key Air Force endeavor.

MARINE CORPS

5-25. The Marine Corps has a culture independent of, but related to, the Navy. Some aspects of the USMC culture include a stress on contingency operations and a unique view of joint operations.

Stress on Contingency Operations

5-26. The Marine Corps is a contingency force. Marines consider themselves an elite force designed to conduct rapid and violent operations. Long-term actions are viewed as resource-draining affairs that are generally ceded to the Army. Stressing the long-term commitment associated with many SF missions is far more likely to generate cooperation with Marine elements than is an attempt to articulate unique capabilities not resident in the Corps.

Unique View of Joint Operations

5-27. The Marine Corps generally is comfortable with joint operations because naval operations incorporate both the Navy and Marine Corps. Given Navy and Marine air assets, Marine ground capability, and Navy maritime capabilities, the Marine Corps is reluctant to integrate beyond the fleet. Marine air assets are rarely ceded to the control of a joint force air component commander (JFACC). Although other Service support is generally accepted, the Marine Corps generally prefers operational concepts supported by organic assets of the combined Navy fleet and the fleet Marine force.

5-28. The recent development of the United States Marine Corps Forces, Special Operations Command (MARSOC), will further enhance the contact that SF Soldiers are likely to have with USMC personnel. Likewise, interoperability between the MARSOC and SF continues to grow, which will facilitate future joint operations.

INTERAGENCY OPERATIONS AND ENVIRONMENT

4-65. Most advisory operations require a high degree of coordination with agencies of the USG, including the Department of State (DOS), USAID, and others. Interagency operations facilitate the implementation of all elements of national power and as a vital link uniting DOD and OGAs. These operations are critical to achieving strategic end states of SO. Interagency operations facilitate unity and consistency of effort, maximize the use of national resources, and reinforce primacy of the political element. A joint HQ conducts interagency coordination and planning. For certain missions, the joint HQ may delegate authority to the component for direct coordination with other agencies.

5-29. In all cases, the component must ensure appropriate authority exists for direct coordination. Components may, in certain special missions, work directly with or for another government agency. In such cases, direct coordination is authorized and command arrangements are specified based on the situation.

5-30. The Office of the Secretary of Defense (OSD) and the joint staff coordinate interagency operations at the strategic level. This coordination establishes the framework for coordination by commanders at the operational and tactical levels. In some cases—such as peacekeeping and complex contingencies—the DOS is the lead agency, and the DOD provides support. In others, DOD is the lead agency.

5-31. The combatant commander (CCDR) is the central point for plans and implementing theater and regional strategies that require interagency coordination. The CCDR may establish an advisory committee to link theater strategy to national policy goals and the objectives of DOS and concerned ambassadors. Military personnel may coordinate with OGAs while operating directly under an ambassador's authority, while working for a security assistance organization, or while assigned to a regional CCDR.

5-32. Coordination between the DOD and OGAs may occur in a Country Team or within a combatant command. Military personnel working in interagency organizations must ensure that the ambassador and CCDR know and approve all programs. Legitimizing authorities determine specific command relationships for each operation. This command arrangement must clearly establish responsibility for the planning and execution of each phase of the operation.

5-33. In addition to extensive USG coordination, commanders must also fully integrate operations into local efforts when appropriate. Such integration requires close coordination with local government agencies and bureaus; local military, paramilitary, or police forces; and multinational partners. A structure such as a mixed military working group composed of senior officials of the military and other agencies may assist such an effort; belligerent parties also may be included.

5-34. The tasks that SF units and individuals execute often evolve from foreign assistance programs. The activities within these programs range from disaster-relief measures to economic and military assistance. It is important, therefore, for SF Soldiers to have an overview of U.S. foreign assistance organizations and collective security agencies and their responsibilities.

AGENDAS

5-35. Each agency has its own organizational agenda, which generally is shaped by U.S. policy. However, interpretations of policies and authorities sometimes vary between government agencies, and sometimes even within the agencies themselves. This may, at times, make achieving unity of effort very difficult. The key to success for the SF advisor is to assess the operational environment, gain an understanding of the interplay between the various agendas and personalities involved, and employ rapport-building and negotiating skills to best support the mission.

INFORMATION

5-36. Each agency has its own approach to problems, its own resources (as well as restrictions on their employment), and its own core values. JP 3-08, *Interagency, Intergovernmental Organization, and Nongovernmental Organization Coordination During Joint Operations (Volume II)*, provides an excellent overview of the characteristics of many groups and agencies. Aside from an understanding of outside organizations, it is important for SF Soldiers to recognize the characteristics of their own organizations, particularly as others may perceive them. This understanding is a critical foundation for communication, interaction, and the decision-making process.

DECISION MAKING

5-37. Most agencies develop operational routines. This makes crisis management very difficult and may hamper even the planning of peacetime activities. Authority for decision-making is usually not delegated to lower levels, making Washington, D.C., rather than the embassy or unified command, the site of many decisions. Under these conditions, developing a set of common objectives for peacetime activities and then unifying their efforts become major challenges.

TRUST

5-38. Because of personal agendas, differences between agencies, and potential organizational friction, establishing and strengthening personal relationships have great significance. Such relationships may be vital to gaining information, cooperation, and overcoming problems.

U.S. FOREIGN ASSISTANCE PROGRAMS

5-39. The majority of U.S. programs for developing nations are economic, political, and humanitarian in nature. Some foreign assistance, however, does take the form of selected military programs. How developing nations resolve their social, economic, political, and military problems influences the prospects for a stable world order. Ultimately, how the problems are resolved impacts—for good or ill—on the security and economic wellbeing of the United States.

5-40. The presence of a U.S. military organization does not determine the level or scope of foreign assistance to individual countries. Nevertheless, the programs discussed below provide the mechanisms through which the United States may render foreign assistance.

DEVELOPMENTAL ASSISTANCE PROGRAMS

5-41. Selected nations receive U.S. developmental assistance primarily for economic and social reasons. This assistance can result in improved security and direct, immediate relief of human suffering. Humanitarian and civic assistance (HCA) helps a nation's development as much as assistance in security matters. HCA is composed of welfare and emergency relief. Developmental assistance programs are administered by USAID.

LOANS

5-42. Developmental loans finance the purchase of a wide range of commodities and related technical services that developing countries need for schools, clinics, irrigation, and roads. The USG may make these loans or they may be offered by private banks with or without government guarantee. Developing countries repay the loans with interest. Interest rates charged to the borrowing country are lower than commercial rates; the United States often approves long-term credit agreements.

TECHNICAL ASSISTANCE

5-43. Technical assistance primarily affects a people's skills, their productivity, and the institutions they build and administer. It allows the people of developing countries to generate what they need for economic and social growth and modernization. Self-sustaining growth depends on the effective use of natural resources, capital facilities, and labor. Technical assistance speeds up the process by which people gain an education, learn skills, and develop positive attitudes so they can more effectively help themselves.

SECURITY ASSISTANCE PROGRAMS

5-44. U.S. nation assistance includes programs that assist friendly foreign countries to establish and maintain an adequate defense posture. These programs also help them to improve internal security and resist external aggression.

5-45. The basis for such assistance lies in the strategy of collective security, a national security policy that recognizes that the security and economic wellbeing of friendly foreign countries are essential to U.S. security. Nation assistance programs aid collective security. They help allied and friendly nations to resist aggression and contribute to national and regional stability.

5-46. Narrowly defined, security assistance is activity pursuant to a body of laws that authorizes and controls the entire process; for example, the Foreign Assistance Act and the Arms Export Control Act of 1976 and related amendments. Considered more properly as a strategic element, security assistance is a tool of U.S. foreign policy. It has application across the spectrum of international competition. It is a bridge that links collective security with U.S. friends and allies in times of peace and in times of crisis. Major security assistance programs include—

- International military education and training (IMET), which includes the following:
  - Formal and informal instruction of foreign students in the United States.
  - Training at civilian institutions.
  - Technical education and training aids.
  - Informational publications.
  - Assistance to foreign military elements by MTTs or technical field training service personnel.
  - Orientation tours of U.S. military installations.
- Foreign military financing (DOD administered).
- Economic support funds (ESFs) (DOS administered).
- Peacekeeping operations (PKOs) (DOS administered).
- Commercial export sales (DOS administered).

Constraints

5-47. Operationally, security assistance is the principal U.S. military instrument for most forms of support to friends and allies. However, its budgetary process makes it largely a long-range preventive tool rather than a short-range reactive tool. The security assistance budget is a part of the DOS (Program 150) foreign assistance budget. The budget-planning cycle takes approximately 2 years to respond to new program requirements. Moreover, the general budgetary climate in which it evolves tends to be extremely limited. Due to these constraints, the United States must usually engage in long-range programs of mutual defense planning with a friend or ally. Specific security assistance initiatives are especially effective in cases where the friend or ally already has a sound financial program for its own defense.

5-48. There are limited, special emergency authorities in the Foreign Assistance Act (FAA) and the Arms Export Control Act (AECA), which the President may use in a crisis to speed up the budgetary process. Nevertheless, they rarely are used and, if used, allow for relatively low levels of USG financing.

5-49. When the United States provides security assistance to a HN, a primary concern is the HN's ability to plan and manage its defense resources by and for itself. HN military organizations may never develop this ability. Additionally, some HNs may continue to request help when it is no longer needed; that is, in areas where they have already achieved self-sufficiency.

Advisory and Training Requirements

5-50. Military advisory and other security assistance personnel need a wide array of skills to handle the diverse activities encompassed in security assistance and FID operations. They need a broad educational foundation to have a better appreciation of the social systems of developing nations. Language training is essential.

5-51. A proper advisor–client relationship depends on successful intercultural communications. Advisors frequently work with counterparts from their respective cultural, educational, and military backgrounds.

5-52. An effective advisor understands the counterpart's sociological, psychological, and political makeup. Accomplishment of the advisory mission often depends more upon positive personal relationships between U.S. advisors and HN counterparts than upon formal agreements. HN leaders may not desire the assistance offered. Indeed, they may tolerate it only to obtain materiel and training assistance. Even when they accept U.S. advice, HN military leaders may not immediately act upon it because of internal constraints and restrictions.

5-53. The U.S. military advisor works in support of an overall U.S. national effort and frequently collaborates in country with civilian members of other U.S. Country Team agencies. Many of the advisor's activities cross mutual jurisdictional boundaries. The advisor must know the functions, responsibilities, and capabilities of the other team agencies. The specific relationship with nonmilitary Country Team members depends largely on the desires of the chief of the diplomatic mission.

Direction and Coordination

5-54. There are a number of agencies involved in the direction and coordination of security assistance activities. These agencies include—

- DOS.
- National Security Council (NSC).
- Central Intelligence Agency (CIA).
- USAID.

5-55. In addition to directing and coordinating security assistance activities, the DOS is also involved in carrying out security assistance activities. Other agencies involved in carrying out these activities include—

- DOD (in particular, the Undersecretary of Defense for Policy).
- Defense Security Assistance Agency (DSAA).
- Joint Chiefs of Staff (JCS).
- Military departments.
- Unified commands.
- Service component commands.
- Security assistance organizations.
- Forces augmenting FID.

5-56. The CCDR appoints a contact officer to represent his interests in each country. The contact officer works with both the diplomatic mission and the HN military forces. This individual heads an organization known as the security assistance office (SAO). The SAO manages DOD security assistance functions in a friendly or allied country and oversees all U.S.-based DOD elements in that country with security assistance responsibilities. Typical SAO designations include military advisory group, MAAG, MILGP, defense field office, Office of Defense Cooperation (ODC), and military liaison office (MLO). In countries where the United States has no SAO, the Defense Attaché may be assigned responsibility for security assistance.

5-57. The SAO is a joint organization. Its chief is essentially responsible to three authorities—the ambassador (who heads the Country Team and controls all U.S. civilian and military personnel in country), the CCDR of the unified command, and the director of the DSAA. The ambassador has operational control

(OPCON) of the SAO for all matters affecting his diplomatic mission, including security assistance programs. Unified Combatant Commands, on the other hand, command and supervise SAOs within their operational theaters in matters that are not the ambassador's responsibility.

5-58. The United States tailors each SAO to the needs of its HN; for this reason, there is no typical or standard SAO organization. However, a large SAO normally has Army, Navy, and Air Force sections. Each of these is responsible for accomplishing its Service portion of DOD security cooperation.

5-59. The SAO can provide limited advisory and training assistance from its own resources. This assistance can, however, be expanded when the SAO is augmented by survey teams, MTTs, ETSS teams, technical assistance field teams (TAFTs), tactical analysis teams (TATs), and other such teams and organizations normally placed under the direction and supervision of the local chief of the U.S. diplomatic mission.

Security Assistance and FID

CCDRs may form area and functional subordinate unified commands. An example of a regional subordinate unified command is United States Forces, Korea (USFK), under the United States Pacific Command (USPACOM). The responsibilities for FID support in these commands closely parallel those discussed for the combatant commands. Specific authority for planning and conducting FID depends on the level of authority delegated by the CCDR. However, basic principles and staff organization remain consistent.

Theater subordinate unified special operations commands (SOCs) are of particular importance because of the significant role of SOF in FID programs. The theater SOC normally has OPCON of all SOF in the theater and has primary responsibility to plan and execute SOF operations in support of FID. SOF assigned to a theater are under the combatant command (command authority) of the geographic combatant commander (GCC). The GCC normally exercises this authority through the commander of the theater SOC. Coordination between the theater subordinate unified SOC and the other component commands of the GCC is essential for effective management of military operations in support of FID, including joint and multinational exercises and MTTs.

EMBASSY ORGANIZATION AND FUNCTION

5-60. SF Soldiers must become familiar with the organization and function of the U.S. embassies in the countries in which they operate. JP 3-07.1, *Joint Tactics, Techniques, and Procedures for Foreign Internal Defense*, provides additional information regarding U.S. embassies.

THE U.S. AMBASSADOR AND THE DIPLOMATIC MISSION

5-61. The permanent U.S. diplomatic mission to a HN is usually called an embassy and includes representatives of all U.S. departments and agencies physically present in the country. The President of the United States (POTUS) appoints and is personally represented by an ambassador who is the head of the embassy. Ambassador is a rank in the U.S. Foreign Service; the official job title of an ambassador in charge of an embassy is chief of mission (COM). Ambassadorial authority extends to all elements of the mission and all official USG activities and establishments within the HN. The POTUS gives the COM direction and control over all U.S. in-country government personnel. However, this authority does not extend to personnel in other missions or those assigned to either an international agency or to a GCC. This last point is a heated topic because—in some circumstances—an ambassador may have control of a CCDR's personnel in-country. He or she can, for example, have individuals declared persona non grata (that is, not welcome) and exclude them from the country. Although the diplomatic mission is outside the

GCC's responsibility, close coordination with each mission in the commander's area of responsibility (AOR) is essential.

5-62. U.S. federal employees detailed to an international organization typically are exempted from routine supervision by the U.S. embassy; however, they still remain under the authority of the COM. For military purposes, the ambassador usually accomplishes supervision either through the assigned SAO or through the Country Team. There is a close coordinating relationship between the ambassador, the represented USG agencies, and the CCDR.

5-63. Section 136 of the 1988–1989 Foreign Relations Authorization Act amended the Foreign Service Act to exclude non-executive-branch (that is, judicial-branch or legislative-branch) employees from COM authority. However, any such persons working in an embassy are usually the subjects of a memorandum of understanding (MOU) placing them under COM authority for everything except purely operational matters.

5-64. The position and authority of the ambassador may vary between peacetime and crisis. During peacetime, or after a return to relatively peaceful conditions, the ambassador usually wields considerable executable control of military activities.

THE COUNTRY TEAM

5-65. The DOS developed the concept of embassy management in the early 1950s, although it wasn't until 1974 that the term Country Team received its first official mention (Public Law 93-475). JP 1-02, *Department of Defense Dictionary of Military and Associated Terms,* defines the Country Team as: "The senior, in-country, U.S. coordinating and supervising body, headed by the chief of the U.S. diplomatic mission, and composed of the senior member of each represented U.S. department or agency, as desired by the chief of the U.S. diplomatic mission."

5-66. Lieutenant Colonel (LTC) Barry K. Simmons offered another description of Country Team in *Executing U.S. Foreign Policy Through the Country Team Concept.* Simmons wrote: "In many ways, the Country Team is a microcosm of what it represents—an assortment of entrenched Washington bureaucratic institutions steeped in the art of turf warfare; self-interest has been known to surface. What tends to prevail in the end, though, is a conviction among the Team's members that they are in fact a team, the Ambassador is the coach calling the plays, and it is their duty to run in the same direction as their teammates. They may seek adjustment at the margins, but they remain on the team and on the field."

5-67. The Country Team's purpose is to unify the coordination and implementation of U.S. national policy within each foreign country under direction of the ambassador. The Country Team advises the Ambassador on matters of interest to the United States and reviews current developments in the country. Other important roles of the Country Team are the identification of potential sources of conflict and threats to U.S. interests in a country and the improvement of condition by introducing programs designed to assist the economy, enhance medical care, and improve the infrastructure of the country.

5-68. The Country Team is the central group responsible for in-country, interdepartmental coordination among key members of the U.S. diplomatic mission or embassy that work directly with the HN government. The members commonly meet at least once a week, usually under the direction of either the ambassador or the deputy chief of mission (DCM).

5-69. The composition of the Country Team makeup is determined by the ambassador and may change from time to time. Subgroups of the Country Team may also be established to deal with special issues (such as counternarcotic [CN] operations). Although the U.S. area military commander (the CCDR or a subordinate) is not a member of the diplomatic mission, the commander may participate or be represented in meetings and coordination conducted by the Country Team.

5-70. Ideally, the Country Team should include representatives of all U.S. departments and agencies present in the country. The U.S. ambassador represents the President, but takes policy guidance from the Secretary of State (SECSTATE) through regional bureaus. The ambassador is responsible for all U.S. activities within the country to which the United States is accredited, and interprets U.S. policies and strategy regarding the HN. The Country Team facilitates interagency action on recommendations from the field and implements effective execution of U.S. programs and policies.

5-71. There are no strict guidelines governing the composition of a Country Team; however, most have similar representation. The Defense Attaché (DATT) and the Chief of the SAO are two principal members. Other common members include the following:

- Ambassador.
- DCM.
- Political counselor.
- Chief of station (COS).
- SAO officers (joint program management section, Army section, Navy section, and Air Force section).
- Director, USAID.
- Regional security officer, to include—
  - Mobile security division (MSD).
  - USMC security guard detachment.
- Consular officer.
- Economic counselor.
- Administrative counselor (the general services officer [GSO] works for the administrative officer and is responsible for buildings, grounds, construction, vehicles, and maintenance).
- Other agency representatives, as desired by the ambassador, to include—
  - Federal Bureau of Investigation (FBI) legal attaché.
  - Civil air attaché.
  - Treasury attaché.
  - Agricultural attaché.
  - Labor attaché.
  - Science attaché.
  - Drug Enforcement Administration (DEA) representative, often referred to as the DEA country attaché (each embassy that has a DEA office is authorized at least two DEA agents).
  - Members of technical assistance teams.
  - Director, Peace Corps.

SPECIAL OPERATIONS SUPPORT TO U.S. AMBASSADORS

5-72. An ambassador or an embassy's Country Team may initiate requests for SOF. The specific request may originate with the ambassador, defense attaché, or military assistance group commander; however, in no case may it occur without the active consent of the concerned ambassador. The requests are passed to the GCC for determination of the appropriate response.

5-73. If the forces are theater-organic and available, and there are no restrictions on their employment (for example, CN operations), the request can be approved by the theater SOC. If there are insufficient forces available in theater, the GCC requests that the Secretary of Defense (SecDef) approve a deployment order for USSOCOM forces through the Chairman of the Joint Chiefs of Staff (CJCS). The Joint Staff ensures that the proper interagency coordination is completed. Once the request is approved and the required coordination is conducted between the DOS, DOD, USSOCOM, and the applicable GCC, specific SOF units or individuals are deployed.

5-74. With few exceptions, SOF deployed to support ambassadors or Country Teams fall under the OPCON of the GCC upon entering the theater. The GCC normally exercises OPCON through the theater SOC, and tactical control (TACON) is given to the SAO. The individual with TACON keeps the ambassador informed of plans and activities during the deployment.

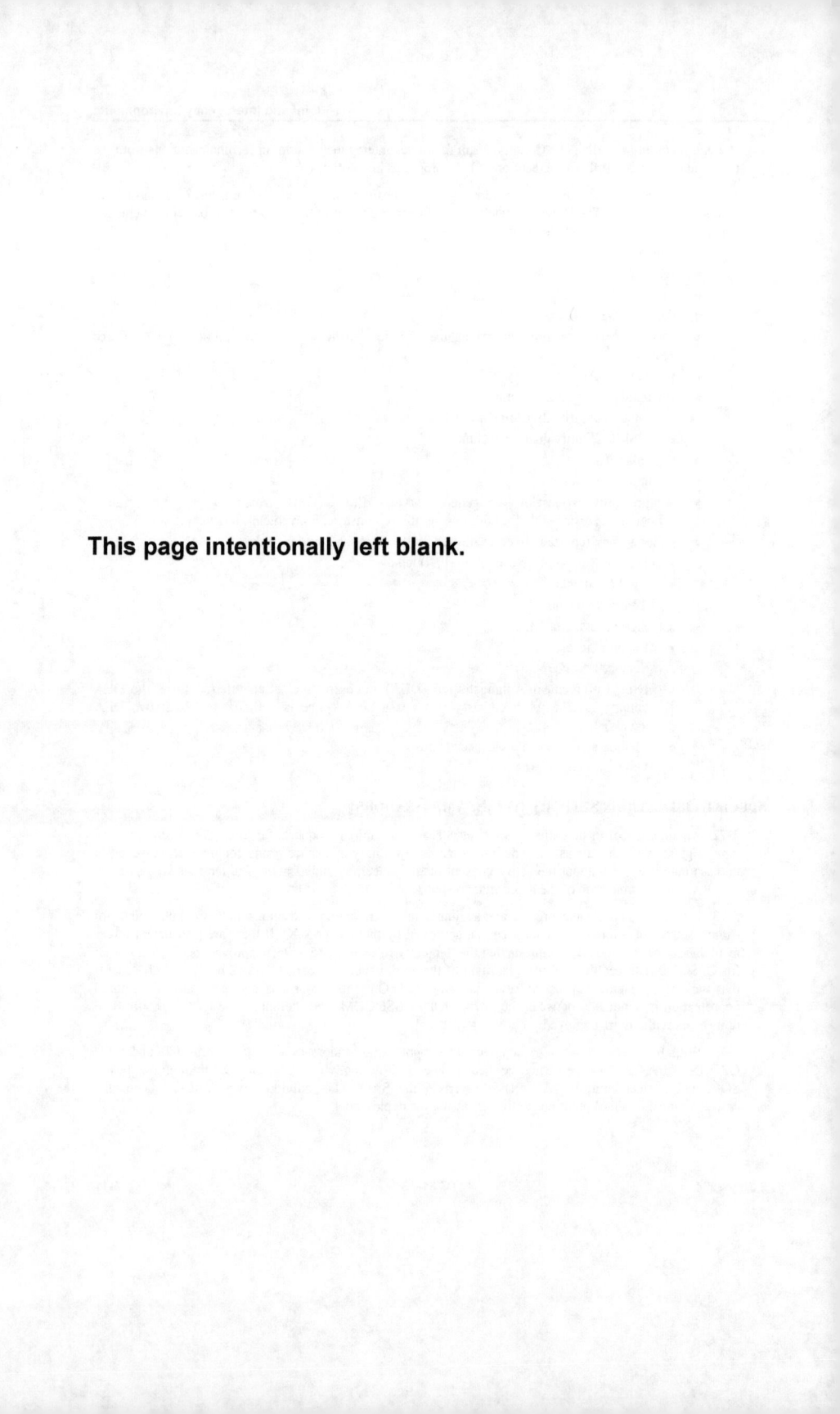

This page intentionally left blank.

Chapter 6

Information Superiority, Information Engagement, and Media Relations

The field of information operations (IO) is vitally important to all aspects of SF operations, and it continues to develop and grow. IO is covered in FM 3-13, *Information Operations: Doctrine, Tactics, Techniques, and Procedures*. The Army's capstone manual for operations—FM 3-0, *Operations*—specifically addresses information superiority and its components, to include information engagement. Public affairs guidance stems from military regulations, including Army Regulation (AR) 360-1, *The Army Public Affairs Program*; Department of the Army Pamphlet (DA Pam) 360-512, *Code of the U.S. Fighting Force*; and DA Pam 360-544, *You and the Law Overseas*. FM 3-61.1, *Public Affairs Tactics, Techniques and Procedures*, is a useful tool for the SF Advisor to use to assist the counterpart's planning for public affairs. Every Special Forces Group (Airborne) has a public affairs office (PAO). GCCs and theater SOCs also have PAOs; these offices can provide public affairs guidance on request. PAOs are capable of providing the advisor with guidance on issues, statements, and media interaction.

INFORMATION SUPERIORITY

6-1. Twenty-first century conflicts occur in operational environments of instant communications. Information systems are everywhere, exposure to news and opinion media is pervasive, the pace of change is increasing, and individual actions can have immediate strategic implications. Information shapes the operational environment at every level.

6-2. Information superiority is the operational advantage derived from the ability to collect, process, and disseminate an uninterrupted flow of information while exploiting or denying an adversary's ability to do the same (JP 3-13). It is imperative that U.S. and allied forces capitalize on the synergy between informational and other operational activities. Therefore, SF advisors must understand information superiority and advise HN counterparts to integrate it into full-spectrum operations as carefully as fires, maneuver, protection, and sustainment. SF advisors must build partner capacity in this critical mission area, promoting informational activity and capability by, with, and through HN forces. Therefore, the SF advisor must ensure HN counterparts can understand, visualize, describe, and direct efforts that contribute to information superiority. As described in FM 3-0, the Army divides these contributors into four primary areas—

- *Army information tasks* are tasks used to shape the operational environment.
- *Intelligence, surveillance, and reconnaissance (ISR) activities* are conducted to develop knowledge about the operational environment. This is an integrated intelligence and operations function, and one which is already thoroughly and historically integrated into SF operations.
- *Knowledge management* is the art of using information to increase knowledge, which enhances situational awareness (SA) at specific times and places and general situational understanding.
- *Information management* is the science of using information systems and methods.

6-3. Of the four primary areas mentioned above, information tasks frequently are the most relevant and challenging tasks that an SF advisor must face on a routine basis, particularly given the twenty-first century's operational environment of instant communications. The Army recognizes five information tasks to shape the operational environment--information engagement, C2 warfare, information protection, operations security (OPSEC), and military deception. The SF advisor needs to prioritize information tasks based on the HN's capabilities and strengths.

6-4.   Within the area of information tasks, SF advisors can have the greatest impact by focusing on the task of information engagement. The basic reason for this is that land operations occur among populations, and as every SF Soldier knows, the key terrain in SF operations is the human terrain. This simple fact requires ground forces to constantly contend with attitudes and perceptions of populations within and beyond their AO. Therefore, SF advisors working with HN counterparts use information engagement in their AOs to communicate information, build trust and confidence, promote support for friendly operations, and influence perceptions and behavior. Information engagement is the integrated employment of several important capabilities. As described in FM 3-0, those capabilities and their intended effects include the following:

- Public affairs (media relations) to inform U.S. and other friendly audiences.
- PSYOP, combat camera, government strategic communications, defense support to public diplomacy, and other means necessary to influence foreign audiences.
- Leader and Soldier engagements to support the efforts listed above.

6-5.   Today's operational environment yields a key and often decisive advantage to the side which best leverages information. As a result, SF advisors need to enable their HN counterparts to provide personal leadership, direction, and attention to fully integrating information. Integrating information tasks into all operations requires including them in the operations process from the inception. This further requires incorporating cultural awareness, relevant social and political factors, and other informational aspects into mission planning and execution. Given this reality, the following information-engagement capabilities deserve particular attention, as the SF advisor can directly influence them:

- First, leader and Soldier engagement is a fundamental and highly effective method to influence people. This involves face-to-face (f2f) interaction by leaders and Soldiers that influences the perceptions of the local populace. In SF, a common aspect of this f2f engagement is frequently referred to as key leader engagement (KLE), a deliberate process in which SF meet with local and regional opinion-shapers. Examples include local tribal elders, elected or appointed officials, military and police commanders, and even NGO leaders.
- Second, combat camera is used to document a wide range of tactical and operational successes and to counter enemy propaganda. If there is no dedicated HN combat-camera capability, the SF advisor should recommend that a HN soldier be designated, equipped, and trained for this function. If that is not possible, the SF advisor should obtain a digital camera and, if necessary, photograph relevant HN activities himself. Positive photographs can have a tremendous effect on information engagement, and digital photos are easily taken and transmitted.
- Third, public affairs (media relations) can have a profound and lasting effect on international perceptions of success or failure of operations. These effects may have strategic consequences, even when dealing with tactical operations. This subject is covered in detail in the next section, beginning with paragraph 6-7.

6-6.   SF advisors must ensure HN counterparts focus their information engagement activities on achieving desired effects locally, however, because land operations always occur in a broader regional and even global context, information-engagement activities must support and complement those of the higher HQ, HN and USG strategic communications guidance (when available), and broader HN and USG policy where applicable.

USE OF THE MEDIA

6-7.   The most difficult portion of information engagement—when necessary—involves dealing with the media. Media contacts normally should be handled by the appropriate PAO. However, this is not always possible, and silence is not always the best solution. Refusal to speak with accredited members of the media may create strong negative impressions with strategic implications.

6-8. When information is withheld, journalists often fall back on speculation. Such speculation is usually inaccurate; however, it is often near enough to the truth that it is accepted by large sections of the public and, in some situations, by established governments. Partisan sources may find it advantageous to leak part of a story to the press in order to build up public support for their position. On occasion, such activities can grow into fully orchestrated press campaigns.

6-9. With modern satellites and communications technology, media are able to distribute reports and photographs faster than the information can be released by the chain of command. Incidents, sometimes fabricated or slanted toward a partisan viewpoint, may be aired in living rooms across the globe the same day. However, this potential liability can be turned into a major advantage for those prepared to deal with the media, as the media provide a powerful and far-reaching opportunity to communicate with critical audiences. Therefore, the SF advisor and HN counterparts frequently are in a position to correct misconceptions and provide accuracy, adequate context, and proper characterization in a timely fashion.

COMMUNICATION WITH THE PRESS

6-10. SF advisors must strive to convey a positive impression and enhance the relationship between the press and the military. Official, published PAO guidance and command policies regarding anonymity and OPSEC must be observed. However, if deemed acceptable within the above guidelines, advisors should attempt to—

- Maintain a list of trusted reporters and editors that cover the operations in the team's AO and keep them informed of significant activities.
- Answer media inquiries promptly, accurately, and courteously. If the answer is not known, the advisor should attempt to find out and get back to the reporter, or refer the reporter to another appropriate source. This may establish the team as a helpful information source and develop a relationship for future balanced coverage.
- Find out reporters' deadlines and use them to the friendly force's advantage.
- Stress the human aspects of a story, the impact of opposing operations on people that readers, viewers, or listeners can identify with. Advisors should point out the needs of the unfortunate, and the fact SF Soldiers and HN counterparts are working to address those needs.
- Encourage the media to see what HN and SF Soldiers are doing and to talk to them about their jobs. Advisors should identify a location in advance for the press to take photos and videotapes.
- Avoid reacting emotionally when reporters or editors appear skeptical or hostile, discuss issues calmly, use facts to back up statements, and maintain focus on the mission.

ADVISOR INTERVIEWS

6-11. What advisors say or do can impact the mission. Consistency with national policy, professionalism, and clear communications are imperative to a successful interview. Advisors should—

- Know and follow the policies of the GCC, U.S. embassy, and PAO regarding media interviews.
- Maintain a general attitude that is friendly, yet professional, and learn and use reporter's names.
- Use language that is clear and easy to understand and avoid military jargon or terminology that others can misinterpret.
- Remain positive, greet the interviewer, and welcome questions. This may be difficult under stressful or tragic conditions; however, a calm, mature appearance earns respect from audiences.
- Prepare notes and study them carefully. Advisors should never walk into an interview unprepared, and should ensure that the facts supporting their position are up to date and come to mind easily.
- Anticipate possible questions and think about various responses. Advisors should rehearse on location if possible and, if time permits, role-play the tough questions.
- Get to know the interviewers. Advisors should research their organizations, the views their organizations possess or try to defend, their agendas, and how they previously conducted interviews.

- Keep the interests of the local nationals or other beneficiaries of the mission at the forefront. Advisors should avoid talking from the point of view of U.S. interests.
- Answer only one question at a time. Advisors should finish the first answer before going to another question if two or more questions are asked at the same time.
- Avoid speculation and give only factual information that can be verified.
- Use caution in quoting statistics as such data may be easily disputed or reinterpreted.
- Be careful when repeating questions. If a question contains incorrect information or inflammatory language, advisors should not repeat it, as this may result in a misquote.
- Strive to tell the truth. If unsure of something, the advisor should admit it. It is better to admit ignorance than mistakenly lie.
- Avoid exaggeration or claims that cannot be backed up.
- Refuse all "off-the-record" discussions, as there is no such thing. Advisors should expect any statement to be quoted.
- Avoid the phrase, "No comment." It makes it appear as if the advisor has something to hide.
- Resist the temptation to attack other groups or organizations, and avoid committing information fratricide (FM 3-13). If questioned about another agency's activities, the advisor should refer to that agency for comment. Advisors should not speak for other organizations.
- Remind reporters that photographing recognizable dead Soldiers, charts, maps, supply depots, or electronic warfare (EW) assets is off limits.
- Remind reporters that their personal security is not a primary military concern.
- Maintain objectivity. Advisors are representatives of the USG and must conduct themselves accordingly. Personal opinions and beliefs should be kept for a more appropriate time and place.

TEAM INTERVIEWS

6-12. Although the PAO normally has responsibility for dealing with the press, it is unlikely that a deployed team will have a PAO representative. Occasions may arise when the HN press will want to question advisors. If conducted properly, these interviews can assist in the mission, and may even improve morale. When preparing team members for an interview, advisors should—

- Identify the team spokesman ahead of time and keep this individual informed.
- Provide team members with a simple theme to convey to the press should they be interviewed.
- Conduct rehearsals in front of video cameras during mock interviews to practice communication of the theme.
- Select interviewees based on their understanding of the theme covered, not their comfort level in front of a reporter. Shyness is a normal trait that may add a human touch to a sensitive situation.

RESPONDING TO MISINFORMATION

6-13. If the media has reported or quoted in an inaccurate and damaging manner, the advisor must use caution in deciding a COA. Advisors inform their chain of command and consider the following questions:

- Is it important enough to correct or is it a detail that—in the long term—is not really important?
- How damaging is the charge, criticism, or error? Will a correction simply give greater visibility to an unfavorable point of view?
- Is a correction worth a restatement of the entire problem to new audiences who did not read or see it the first time? Is it possible to target only the audience originally exposed to the story?
- Can significant gain result from pointing it out?

6-14. When misinformation is released to public through government or media sources, the COA taken by the advisor will be unique to the situation, but it must include a plan to ensure that relevant government officials and the media recognize the inaccuracy of the information. This may prevent future incidents of misinformation. Operational and combat experience since 2001 shows that in the struggle for information superiority it is critical to be candid and honest, and to not allow enemy lies to stand unchallenged, particularly in areas of great international interest.

Appendix A
Advisor Checklists

INITIAL CONTACT

A-1. The predeployment site survey (PDSS) leader—along with any subordinates he may specify—establishes effective initial rapport with the HN unit commander. The PDSS leader—

- Conducts introductions in a businesslike, congenial manner using the target-country language.
- Briefs the HN commander on the SF advisors' PDSS mission and the restrictions and limitations imposed on the detachment by the higher U.S. commander. The PDSS leader should use the target-country language and, if required, visual aids translated into the HN language.
- Assures the HN commander that all PDSS team members are fully supportive of the HN's position and that they firmly believe a joint SF and HN-unit effort will be successful.
- Assures the HN commander that his assistance is needed to develop the tentative objectives for advisory assistance.
- Deduces or solicits the HN commander's actual estimate of his unit's capabilities and perceived advisory assistance and material requirements.

Note: The PDSS leader should not make any promises or statements that could be construed as promises to the HN commander regarding commitments to provide the advisory assistance or fulfill material requirements.

- Explains the PDSS team's initial plan for establishing counterpart relationships, obtains approval from the HN commander for the plan, and requests to conduct the counterpart linkup under the mutual supervision of the PDSS leader and the HN commander.
- Supervises the linkup between PDSS team members and their HN counterparts to determine if the HN personnel understand the purpose of the counterpart relationship and their responsibilities within it.

A-2. The PDSS team members analyze the HN unit's status IAW their functional area for the purpose of determining the HN requirements for advisory assistance. The PDSS team members—

- Explain the purpose of the analysis to counterparts.
- Encourage counterparts to assist in the analysis, the preparation of estimates, and the briefing of the analysis to the SFODA and HN unit commanders.
- Collect sufficient information to confirm the validity of current intelligence and tentative advisory assistance COAs selected prior to deployment.
- Collect and analyze all information relating to FP.
- Prepare written, prioritized estimates for advisory assistance COAs.
- Brief, with their counterparts, the estimates to the PDSS team and HN unit commander.
- Inspect, with their counterparts, the HN facilities that will be used during the assistance mission IAW their functional area and the SFODA operation order (OPORD).
- Identify deficiencies in the facilities that will prevent execution of the tentatively selected advisory assistance COAs.
- Prepare written or verbal estimates of COAs that will correct the deficiencies or negate their effects on the tentatively selected advisory assistance COAs.
- Supervise the preparation of the facilities and inform the SFODA commander of the status of the preparations compared to the plans for them.

A-3. Once received, the PDSS leader supervises the processing of the survey results. The PDSS leader then—

- Recommends to the HN unit commander the most desirable COAs emphasizing how they satisfy actual conditions and will achieve the desired advisory assistance objectives.
- Ensures that his counterpart understands that the desired COAs are still tentative contingent on the tasking U.S. commander's decision.
- Selects the COAs to be recommended to the follow-on SF units, after obtaining input from the HN unit commander.
- Ensures the higher in-country U.S. commander is informed of significant findings in the team survey for HN assistance.

A-4. The PDSS team plans its security IAW the anticipated threat. Adjustments are made as required by the situation on the ground. The PDSS team members—

- Fortify their positions (quarters, communications, medical, command) IAW the available means and requirements to maintain low visibility.
- Maintain a team internal guard system, aware of the locations of all other SF advisors, and ready to react to an emergency by following the alert plan and starting defensive actions.
- Maintain a team internal alert plan that will notify all team members of an emergency.
- Maintain communications with all subordinate team members deployed outside of the immediate area controlled by the team.
- Establish plans for immediate team defensive actions in the event of an insurgent or terrorist attack or a loss of HN rapport with hostile reaction.
- Discuss visible team security measures with HN counterparts to ensure their understanding and to maintain effective rapport.
- Encourage the HN unit, through counterparts, to adopt additional security measures that have been identified as necessary during the analysis of the HN unit status and the inspection of its facilities.
- Establish mutual plans with the HN unit, through counterparts, for defensive actions in the event of an insurgent or terrorist attack.
- Rehearse team alert and defensive plans.
- Encourage the HN unit, through counterparts, to conduct mutual, full-force rehearsals of defensive plans.

EVALUATION

A-5. SF staff sections identify additional training objectives or necessary modifications to HN operating procedures IAW their functional area. These personnel—

- Consolidate training reports, AARs, and mission debriefings to produce a list of deficiencies while avoiding redundancy or closely related items.
- Identify HN personnel in key positions who require additional leadership or other functional duty training.
- Review the status of the threat to determine its impact on HN performance.
- Brief counterparts on the identified deficiencies and the threat impact to establish their understanding.
- Encourage counterparts to participate in the identification of additional training objectives or modifications to HN operating procedures.
- Analyze the required training objectives or procedure modifications and the threat status to develop estimates of tentative COAs that can meet the requirements.
- Identify new tasks specified or implied by the higher U.S. commander.

A-6. The SF staff sections develop a new program of instruction (POI) or make the necessary modifications to HN procedures. The senior SF advisor selects the most desirable COAs and encourages his counterpart to approve them. Additionally, the staff and senior advisor—

- Develop plans for selected COAs that are based on appropriately modified U.S. doctrine and contain all necessary annexes, schedules, and lesson outlines.
- Develop plans for selected COAs that reflect a logical progression from the present deficient status to the desired improved status.
- Encourage counterparts to become involved in the development of the plans to improve their self-sufficiency.
- Review the plans with their counterparts to ensure they are satisfactory to the SF advisor and the HN unit.
- Identify any necessary deviations from the mission guidance issued to the SF advisor by the higher commander.
- Identify any necessary additional resources and the supporting section or unit that can provide them.
- Emphasize human rights training as required.
- Emphasize, as appropriate, multi-echelon HN training in the new or revised POI by planning to teach individual, crew, leader, and collective skills concurrently.

A-7. The SFODA prepares to execute the newly developed advisory assistance plans by—

- Informing the higher U.S. commander of the plans and obtaining approval for their execution.
- Informing the higher U.S. commander of irresolvable HN personnel difficulties.
- Ordering execution of the plans after obtaining an agreement from the counterpart.
- Requesting necessary additional support.
- Preparing in-country facilities, as required.
- Requesting additional language support, as required.
- Encouraging the participation of counterparts in the preparation of the plans.

ADVISING

A-8. The senior SF advisor—typically a Special Forces officer (MOS 18A) detachment commander—assists the HN unit commander in beginning the C2 process. The SF advisors and staff members assist their HN counterparts in the development of the COIN operation plan (OPLAN) or OPORD IAW their functional area. These individuals—

- Review the estimates of tentative COAs developed by their counterparts and recommend improvements or additional COAs to satisfy the SF advisor and HN unit commander's planning guidance.
- Recommend to the senior SF advisor the most desirable COAs.
- Develop, unilaterally, contingency plans for gaps in HN unit planning if their counterparts are unreceptive to recommendations for improvements.
- Review for completeness the portions of the OPLAN or OPORD prepared by their counterparts and recommend improvements or additions needed to satisfy the SF advisor team and HN unit commanders' planning guidance.
- Encourage their counterparts to complete, as quickly as possible, their planning tasks and to adhere to the event time plan.
- Keep the senior SF advisor apprised of the status of the planning process.
- Offer additional assistance as required.

SPECIAL FORCES OFFICER (MOS 18A)

A-9. The 18A commands the detachment and advises his HN counterpart. The 18A—

- Accompanies the HN unit commander when he receives his mission or monitors the situation to assist him in deducing or anticipating his next mission in the earliest time possible.
- Monitors the HN unit commander's identification of the objective, his higher commander's intent, and all specified or implied tasks, and recommends improvements or additions, as needed.
- Reviews for completeness the HN unit commander's selection of the essential tasks that must be planned for, and recommends improvements or additions.
- Reviews the HN unit commander's identified operational constraints and restraints for completeness, and recommends improvements and additions IAW appropriately adapted U.S. doctrine, the higher U.S. commander's guidance, and applicable U.S. and HN operational agreements.
- Reviews the HN unit commander's event time plan, and recommends improvements using the one-third to two-thirds rule.
- Encourages the HN unit commander to brief his staff on the mission and issue his planning guidance as soon as possible.
- Briefs the SF advisor staff on the mission, issues, and planning guidance as early as possible.

SPECIAL FORCES WARRANT OFFICER (MOS 180A)

A-10. The 180A advises his HN counterpart in his functional duties. The 180A—

- Monitors all HN staff sections and recommends changes in organization and procedures, as necessary, to improve efficiency.
- Assists the counterpart during periods when he is in command of the HN force (in the absence of the commander).
- Assists counterparts in the preparation of all orders and plans.
- Assists in the planning, coordination, and implementation of PSYOP and CA tasks assigned to the HN unit or determined to be desirable.
- Monitors liaison and coordination with higher HN HQ and recommends changes, as necessary, to improve efficiency.
- Informs the senior SF advisor of any significant problems identified and provides recommendations for rectifying them.
- Reviews reports on human-rights violations and forwards these reports through the SF advisor chain of command.

SPECIAL FORCES OPERATIONS SERGEANT (MOS 18Z)

A-11. The 18Z advises his HN counterpart in his functional area duties. The 18Z—

- Assists in the use of estimates, predictions, and information provided by the intelligence staff section (S-2) in the preparation of tactical plans.
- Assists the 180A in the preparation of all orders and plans.
- Assists in the supervision of training and preparations for operations.
- Makes recommendations to ensure that operations remain consistent with overall goals.
- Informs the senior SF advisor of any significant problems identified and provides recommendations for rectifying them.

SPECIAL FORCES INTELLIGENCE SERGEANT (MOS 18F)

A-12. The 18F advises his HN counterpart in his functional area duties. The 18F—

- Monitors the situation map and recommends actions to keep it current based on the available intelligence, as necessary.

- Monitors the collection, interpretation, and dissemination of information concerning the effects of weather, terrain, and insurgent forces on the HN unit's mission and recommends improvements in procedure, as necessary.
- Manages the submission and receipt of intelligence reports, as necessary, to ensure that all assets are exploited.
- Provides intelligence-collection plan instruction and advice to HN counterparts.
- Recommends improvements, as necessary, to the HN tactical operation center (TOC) communications SOP to ensure that the S-2 receives SITREPs from the operations staff section (S-3), the fire direction center (FDC), and all unit attachments.
- Assists in the evaluation and interpretation of intelligence information to determine insurgent or terrorist probable COAs.
- Monitors the dissemination of intelligence information to the HN commander, staff, higher HQ, subordinate units, and attachments, and recommends improvements, as necessary.
- Assists the 18Z and his counterpart in supervising and controlling reconnaissance and surveillance activities.
- Assists in the briefing and debriefing of patrols operating as part of reconnaissance and surveillance activities.
- Assists the 18Z and his counterpart in the development of plans for reconnaissance and surveillance activities to ensure the most complete coverage of the AO.
- Assists in the interrogation of EPWs and detainees.
- Coordinates with the HN S-3 to adjust reconnaissance and surveillance plans IAW changes in the situation.
- Assists in originating tasking requests for intelligence support to ensure all available assets are exploited.
- Monitors procedures used to protect classified and operationally sensitive material and recommends improvements, as necessary.
- Continuously updates the intelligence preparation of the battlefield (IPB) prepared during predeployment.
- Informs the senior SF advisor of any significant problems identified and provides recommendations for rectifying them.

SPECIAL FORCES MEDICAL SERGEANT (MOS 18D)

A-13. The 18D advises his HN counterpart in personnel staff officer (S-1) functional area duties. The 18D—

- Monitors the maintenance of HN unit strength and recommends improvements, as necessary.
- Monitors the processing of HN wounded and personnel killed in action (KIA) and recommends improvements, as necessary.
- Monitors the processing of EPWs and detainees, recommending that they be handled IAW the 5 Ss—silence, search, segregate, secure, and safeguard.
- Monitors HN unit morale and recommends actions to improve it, as necessary, IAW HN custom.
- Monitors HN unit discipline and maintenance of order and recommends actions to improve it, as necessary, IAW HN custom.
- Informs the senior SF advisor of any significant problems identified and provides recommendations for rectifying them.

SPECIAL FORCES ENGINEER SERGEANT (MOS 18C)

A-14. The 18C advises their HN counterpart in his functional area duties. The 18C—

- Monitors the maintenance of equipment readiness, recommending improvements, as necessary.
- Monitors the support provided in all classes of supply to the HN unit, its subordinate units, and attachments, and recommends improvements, as necessary.

- Assists in the supervision of the use of transportation assists.
- Informs the senior SF advisor of any significant problems identified and provides recommendations for rectifying them.

SPECIAL FORCES WEAPONS SERGEANT (MOS 18B)

A-15. The 18B advises the HN unit fire support officer (FSO) in his functional area duties. The 18B—

- Assists in the planning, coordination, and request for fire support for the HN unit and the employment of its fire-support assets.
- Makes recommendations to ensure that fire support is employed IAW firepower restrictions and the principle of minimum essential force.
- Monitors communications procedures for requesting fire support and recommends improvements, as necessary, to improve efficiency.
- Assists in the processing of fire-support requests to ensure the timeliness and accuracy of the response.
- Informs the senior SF advisor of any significant problems identified and provides recommendations for rectifying them.

SPECIAL FORCES COMMUNICATIONS SERGEANT (MOS 18E)

A-16. The 18E advises their HN counterpart in his functional area duties. The 18E—

- Monitors the use of communications nets.
- Monitors the maintenance of communication equipment and recommends improvements, as necessary.
- Recommends improvements to signal plans so that available communications assets may be exploited to gain every possible advantage.
- Informs the senior SF advisor of any significant problems identified and provides recommendations for rectifying them.

EXECUTION

A-17. The senior SF advisor and staff members review the HN unit OPLAN or OPORD. These individuals—

- Recommend improvements to the task organization in order to maximize the strengths and minimize the weaknesses of the available assets, counter the anticipated threat, allow for swift transitions in the organization for contingencies, and maintain a reserve appropriate to the size of the HN force employed.
- Recommend improvements to the intelligence portions of the plan or order so that all relevant sources of information have been exploited and the information has been analyzed to allow the HN commander to plan actions that seize the initiative.
- Recommend improvements to the execution so that the mobility of the HN unit is employed to achieve all possible tactical advantage, the minimum firepower needed is used to accomplish the given tasks, restrictions on the rules of engagement are specified, and subunit missions (to include the reserves) are clear.
- Recommend improvements to the service and support plans so that only mission-essential supplies and equipment are taken, that soldiers are not overburdened at the expense of the mobility, and that resupply and medical evacuation (MEDEVAC) are available as needed.
- Recommend any other improvements needed to ensure the OPLAN or OPORD is complete, to include human rights considerations, measures, or required training.

DISSEMINATING THE PLAN

A-18. The SFODA members monitor the dissemination of the HN unit's OPLAN or OPORD and mission preparations by their counterparts. The SFODA members—

- Attend the issuance of the OPLAN or OPORD and recommend additions or clarifications to the verbal plan or order needed for completeness and understanding.
- Encourage counterparts to inspect weapons and ammunition, vehicles, and mission-essential supplies and equipment, and recommend actions needed to correct shortcomings IAW the OPLAN or OPORD, or newly identified requirements.
- Attend mission rehearsals (brief back, reduced force, or full force) and recommend any required additions or modifications needed to cover actions in the objective area, actions on enemy contact, and alternate COAs for reasonable contingencies.
- Report all identified HN unit planning or preparation deficiencies to the senior SF advisor and, IAW functional areas of responsibility, the applicable SFODA member.

PREPARING FOR EXECUTION

A-19. The SFODA members prepare for participation in the operation. They accomplish this by—

- Ensuring primary, alternate, and emergency communications between SFODA elements are established and functioning IAW the resources available.
- Disseminating unilateral contingency plans to all SFODA members and rehearsing them.
- Discussing the identified planning or preparation deficiencies with the HN unit commander and attempting to resolve them.

Note: The SFODA commander may withhold specific portions of SFODA assistance that would place SFODA members at personal risk due to unacceptable conditions resulting from uncorrected planning or preparation deficiencies.

- Submitting premission reports to the next-higher U.S. commander IAW requirements in the SFODA OPORD.

EXECUTING THE MISSION

A-20. The senior SF advisor assists the HN unit commander in providing C2 during the execution of the operation. The senior advisor—

- Monitors the tactical situation and recommends changes to the present COA to gainfully exploit changes in the situation.
- Monitors the location of the HN commander and recommends changes so that he can provide leadership at critical points and not deprive himself of the ability to maneuver his force in response to tactical changes.
- Monitors the information flow to the HN commander and recommends improvements needed to—
  - Make continuous use of intelligence-collection assets.
  - Keep subordinates reporting combat information.
  - Screen the information given to the HN commander to prevent information overload.
  - Keep the command communications channels open for critical information.
- Monitors the HN commander's control of the execution and recommends improvements to—
  - Focus combat power on the objective.
  - Keep movement supported by direct and indirect fire.
  - Maintain mutual support between subordinate elements.

- Maintain fire control and discipline.
- Consolidate and reorganize during lulls in the battle or after seizing the objective.
- Monitors any command succession and assists the new HN unit commander to smoothly and rapidly take control of the execution of the operation.

A-21. The SFODA members also assist their counterparts during the execution of the operation. The SFODA members—

- Monitor staff functions IAW their functional area and recommend improvements or corrections, as needed.
- Monitor the technical or tactical execution of individual tasks and recommend improvements or corrections, as needed.
- Remain continuously aware of the tactical situation.
- Execute SFODA unilateral contingency plans, as required by the situation.
- Note reoccurring or significant problems or events for reference during end-of-mission debriefings and reports.

CONDUCTING END-OF-MISSION ACTIVITIES

A-22. Upon completion of the mission, the SFODA conducts end-of-mission activities. SFODA members—

- Participate in HN unit debriefings, encouraging the HN unit commander and important subordinates to realistically appraise the HN unit's performance and to modify their TTP to improve future performance.
- Conduct a unilateral SFODA debriefing to identify reoccurring or significant problems for both the HN unit and the SFODA.
- Modify the SFODA's mission execution plan to correct identified problems.
- Make recommendations for awards for both HN and SFODA personnel, as applicable.
- Document and report to the higher U.S. commander—
  - Incidents of corruption.
  - Gross inefficiency.
  - Violations of human rights.
  - Actions of HN military or government officials who habitually hinder operations through incompetence, self-interest, or suspected sympathy for the insurgents or terrorist.

Note: Reports to the higher U.S. commander regarding such deficiencies are very sensitive and must be monitored closely to ensure complete security.

TRAINING

A-23. The SFODA conducts training and executes a POI. In order to accomplish this, the SFODA—

- Adheres to the training schedule consistent with cooperation from the HN forces and changes in the mission, enemy, terrain and weather, troops and support available, time available, and civil considerations (METT-TC).
- Encourages, through counterparts, HN unit commanders to ensure all their personnel receive training as scheduled.
- Rehearses all classes with counterparts and, as necessary, with interpreters.
- Ensures all training objectives satisfy actual HN training needs identified during the analysis of the HN units' status (unless ordered to do otherwise by the higher U.S. commander).
- Ensures all training objectives are structured IAW applicable U.S. military doctrine unless specific modifications to doctrine are made to meet an identified in-country need.
- Implements multi-echelon training by teaching individual, crew, leader, and collective skills concurrently.

A-24. The SFODA presents the instruction. SFODA members—

- Adhere to the lesson outlines consistent with the cooperation from the HN forces and changes in the METT-TC.
- State clearly the task, conditions, and standards to be achieved during each lesson at the beginning of the training (to include training exercises) ensure the HN students understand them. (Human rights should be emphasized in the appropriate period of instruction.)
- Demonstrate the execution or show the desired end result to clearly illustrate the task.
- Stress the execution of the task as a step-by-step process, when possible.
- Monitor the HN students' progress during practice and correct mistakes as they are observed.
- State (at a minimum) all applicable warning and safety instructions in the HN language (without the use of an interpreter).
- Monitor periodically instructions given through HN interpreters to ensure accurate translations using a HN-language-qualified SFODA member.

A-25. Designated SFODA members maintain written administrative training records. These members—

- Encourage HN counterparts to assist.
- Record all HN personnel and units who receive training and identify the type of training they receive.
- Organize records to identify training deficiencies and overall level of HN proficiency.
- Identify specific HN personnel or units who demonstrate noteworthy (good or bad) performance.
- Identify to the SFODA and HN unit commanders the noted training deficiencies, noteworthy performances, and required additional or remedial training.

A-26. The SFODA conducts AARs after all collective HN unit training events. SFODA members—

- Develop a discussion outline to guide the AAR.
- Review the training objectives with the concerned HN unit commander by asking leading questions, surfacing important tactical lessons, exploring alternate COAs, keeping to teaching points, and making the AAR positive.
- Encourage the concerned HN unit commander to review the training event with his entire unit (or key subordinate leaders, as applicable) by stressing how he will strengthen his chain of command and put focus on himself as the primary trainer of his unit.
- Stress to the HN unit commander the importance of discussing in his review not only what happened, but also why it happened; the important tactical lessons learned; alternate COAs that could have been taken; and important teaching points.
- Avoid criticizing or embarrassing the HN unit commander.
- Monitor the HN unit commander's review of the training with his unit to ensure the focus is on the training objectives and the lessons learned.
- Prepare a report of the evaluation of the HN unit and forward it to the staff section maintaining the administrative training records.

A-27. The SFODA ensures the security of the training sites. SFODA members—

- Analyze the threat to determine any capabilities to attack or collect intelligence on the HN unit's training at each site.
- Prepare estimates of COAs that would deny the training sites to the insurgents or terrorist.
- Recommend to the HN unit commander that he order the adoption of the most desirable COA, stressing how it best satisfies the identified need.
- Ensure before each training session (using, as a minimum, brief back rehearsal) that all personnel—both U.S. and HN—understand the defensive actions to be taken in the event of an insurgent or terrorist attack and any OPSEC measures to be executed.

ONGOING ACTIVITIES

A-28. The SFODA members maintain their functional area portion of an advisor database (information files) IAW unit SOP. SFODA members—

- Request information necessary to satisfy the commander's critical information requirements (CCIR) from applicable sources.
- Route information requests IAW unit SOP through SF advisor to higher U.S. HQ.
- Route intelligence information requests to the S-2.
- Identify information received that satisfies CCIR which concern them.
- Modify previously developed estimates IAW the latest information available.
- Notify other concerned staff sections of modified estimates and plans.
- Notify other concerned (higher, lower, or adjacent) staff sections of information, as it is identified, that satisfies their information requirements (IRs).
- Update the SF advisors CCIR list IAW the latest information available and requirements for additional CCIR that arise from modified estimates and plans.

A-29. In addition to the responsibilities listed above, the 18F executes a number of additional duties. The 18F—

- Updates the IPB prepared in predeployment.
- Supervises the dissemination of intelligence and other operationally pertinent information within the SF advisor team and, as applicable, to higher, lower, or adjacent concerned units or agencies.
- Monitors the implementation of the SF advisor team intelligence collection plans to include the update of the SF advisor priority intelligence requirements (PIRs) or IRs, conducting area assessment and coordinating for additional intelligence support.

PSYCHOLOGICAL OPERATIONS AND CIVIL AFFAIRS

A-30. The 180A supervises mission analysis and determines PSYOP requirements. The 180A—

- Assesses the psychological impact of the SFODA's presence, activities, and operations in the AO.
- Reviews the OPLAN or OPORD to ensure it supports U.S. and HN psychological objectives.
- Coordinates the analysis of each of the detachment's official duties to determine their psychological effect.
- Considers the psychological impact on the populace of SFODA participation in such events as military ceremonies, religious services, and social events when deciding if the SFODA should participate.
- Determines the psychological effects of training during periods of national holidays or religious holidays and schedules IAW this determination.

A-31. SFODA members should conduct themselves in a manner that takes into account local customs and traditions as well as Army standards of conduct. The SFODA commander—

- Ensures that all detachment members respect HN and local customs, courtesies, and taboos, and conduct themselves in a correct and professional manner.
- Emphasizes that, as members of an SFODA and representatives of the United States to the HN, any action taken by an American—good or bad, on duty or off duty—will have a psychological impact on the mission.
- Monitors and corrects discrepancies.

A-32. Planned PSYOP activities are integrated into each SFODA operation in order to establish a favorable U.S. image in the HN and further accomplishment of the SFOD mission. SFODA members—

- Coordinate with trained PSYOP assets to capitalize on positive mission successes.
- Coordinate with U.S. agencies to facilitate the use of HN and commercial media assets that influence personnel in the AO.

- Emphasize U.S. support of HN programs (not U.S. unilateral operations) in all PSYOP products and on all operations.
- Incorporate PSYOP products and activities that portray a positive U.S. and HN image in each SFODA activity.

A-33. The SFODA advises and assists HN forces in gaining or retaining the support of the local populace, discrediting the insurgents, and isolating the insurgents from the populace. SFODA members accomplish this by—

- Influencing HN forces through advice and example to conduct themselves IAW acceptable military norms, ethics, and professionalism, to include the principles of leadership and standards of conduct.
- Training HN leadership in the advantages and techniques of maximizing public opinion in favor of the HN and SFODA mission and to the discredit of the insurgents.
- Coordinating for close and continuous PSYOP support to maximize the effect of CA operations. This includes advising the HN to utilize its own resources in the same manner.
- Integrating PSYOP capabilities into PRC measures to disseminate information and explains the rationale for the program.

CIVIL-MILITARY OPERATIONS

A-34. If CA and civil-military operations (CMO) are a significant part of the mission, CA personnel may be attached to the SFODA. If no CA personnel are attached to the SFODA, the 180A usually performs these duties.

A-35. The SFODA should establish contact and attempt to coordinate with appropriate nonmilitary agencies of the HN and the U.S. mission, consider synchronization of its military operations with the programs of these agencies, and advise supported HN forces on integrating CMO into their operations. The 180A supervises mission analysis to determine CMO requirements and—

- Determines the political, economic, social, and cultural factors that influence SFODA operations in the AO.
- Determines the security needs of the SFODA and of the local population in the AO.
- Requests and reviews the internal development objectives, policies, plans and programs of the HN and U.S. mission from the next higher HQ.
- Includes the CA estimate in the military decision-making process.
- Requests any CA support required to conduct the mission.
- Coordinates operations with appropriate HN, U.S. mission, and international agencies.
- Defines CA mission and CMO tasks.
- Requests and reviews USAID 1- and 5-year plans.
- Requests from the next higher HQ information on CA activities conducted in the AO by other agencies.
- Conducts a postinfiltration area assessment to validate and update CMO-related information in the area study. Incorporates changes and additions in the area assessment and modifies plans and operations to account for these changes.

A-36. The SFODA provides civil assistance to HN government agencies. SFODA personnel conduct CMO based on the CA annex to the OPORD, supervise attached or assigned CA personnel, and direct and support assigned or attached CA units of detachment to company size. The SFODA establishes contact with local governments within the AO (and advises HN forces to do the same), takes actions intended to establish and maintain favorable relationships with the local population and the recognized government, and uses civil communications available in the AO to disseminate civil information.

CIVIL DEFENSE

A-37. The SFODA may advise and assist HN forces in planning and implementing a civil defense program. SFODA members—

- Analyze the civil defense structure to ensure that it meets identified security needs.
- Assess civil defense planning for the presence and effectiveness of emergency welfare services and emergency food, water, sanitation, and medical supplies.
- Coordinate civil defense activities of fire, police, and rescue personnel with those of the military to achieve unity of effort.
- Identify civilian evacuation plans and assesses their adequacy.

DISPLACED PERSONS

A-38. The SFODA advises and assists HN forces supporting displaced person operations. SFODA members advise and assist the HN in—

- Estimating the number of displaced civilians, their points of origin, and their anticipated direction of movement.
- Planning movement control measures, emergency planning, and evacuation of dislocated civilians.
- Coordinating with military forces for transportation, military police (MP) support, military intelligence (MI) screening and interrogation, and medical activities, as needed.
- Establishing and supervising the operation of temporary or semipermanent camps for displaced civilians.
- Resettling or returning displaced civilians to their homes in accordance with U.S. and HN policy and objectives.
- Establishing camps and relief measures for displaced civilians.
- Monitoring the conduct of movement plans for displaced of civilians.

POPULACE AND RESOURCES CONTROL

A-39. The SFODA advises and assists HN forces and agencies in the planning and implementation of PRC programs. SFODA members—

- Identify PRC requirements.
- Assist in planning and coordinating PRC measures that meet these requirements.
- Integrate PRC measures with PSYOP to obtain popular acceptance and support of the measures.
- Provide advice and assistance indirectly to minimize direct U.S. involvement and emphasize low-visibility U.S. support of HN programs.
- Evaluate adequacy of PRC programs and recommend improvements, as required.

A-40. The SFODA identifies and acquires HN resources to assist the SFODA in mission execution. SFODA members should determine the political organizations and key leaders existing in the AO and surrounding country to facilitate gaining civilian cooperation. SFODA members advise and assist the HN to minimize civilian interference with tactical operations and to help the HN—

- Anticipate civilian reactions to planned military operations and plans to accommodate that reaction.
- Provide aid that will improve conditions for civilians who are destitute and reduce theft and destruction of both military and indigenous property.
- Determine methods and techniques of operation that will be most acceptable to the populace and still allow for the accomplishment of the SFODA mission.
- Identify military COAs to avoid civilian population centers and rural activities when feasible.
- Coordinate PRC measures to remove civilians from probable battle areas with HN military and civil authorities.

A-41. The SFODA must strive to meet legal and moral obligations to the local populace and the families of supporting HN forces. SFODA members—

- Observe laws of armed conflict and ROE.
- Report human rights violations by HN forces or insurgent forces to higher HQ.
- Act promptly to prevent or stop human rights violations (within capabilities).
- Establish medical treatment programs on a space-available basis (within capabilities).
- Provide emergency disaster relief in a life-threatening situation (within capabilities).

A-42. The SFODA advises and assists the HN in protecting cultural properties in the AO. SFODA members should locate and identify religious buildings, shrines, and consecrated places. Additionally, they should recommend and observe ROE that protect these sites during military operations.

INTEGRATING PSYCHOLOGICAL OPERATIONS AND CIVIL-MILITARY OPERATIONS

A-43. The 180A coordinates and integrates CMO with PSYOP. The 180A—

- Advises and assists the HN to ensure the populace is informed of tactical victories and U.S. and HN civic-action efforts in their benefit, and coordinates available U.S. support to HN forces to accomplish this aim.
- Advises HN to reduce PRC when the enemies of the people are denied support and supplies.

This page intentionally left blank.

Appendix B
Cultural Checklists

CULTURAL

B-1. The SF advisor must conduct a thorough review of the target country and culture. Figure B-1, pages B-1 through B-3, provides 50 sample questions that the SF advisor should consider when conducting his review. It is not an all-inclusive list; it is a starting point, prodding the advisor to consider other questions.

---

1. Who are the people who are prominent in the affairs (that is, military, politics, athletics, religion, and the arts) of the host country?

2. Who are the country's national heroes and heroines?

3. What is the national anthem and how does it sound?

4. Are other languages spoken besides the dominant language? What are the social and political implications of language usage?

5. What is the predominant religion? Is it a state religion? What are its sacred writings and have they been read or reviewed?

6. What are the most important religious observances and ceremonies? How regularly do people participate in them?

7. How do members of the predominant religion feel about other religions?

8. What are the most common forms of marriage ceremonies and celebrations?

9. What is the attitude toward divorce? Extramarital relations? Plural marriage?

10. What is the attitude toward gambling?

11. What is the attitude toward drinking?

12. Are the prices asked for merchandise fixed, or are customers expected to bargain? How is bargaining conducted?

13. If, as a customer, a person touches or handles merchandise for sale, will the storekeeper think they are knowledgeable, inconsiderate, within their rights, completely outside their rights? Other?

14. How do people organize their daily activities? What is the normal meal schedule? Is there a daytime rest period? What is the customary time for visiting friends?

Figure B-1. Cultural analysis questionnaire

15. What foods are most popular and how are they prepared?

16. What things are taboos in this society?

17. What is the usual dress for women? For men? Are slacks or shorts worn? If so, on what occasions? Do teenagers wear jeans?

18. Do barbers and hairdressers use techniques similar to those used by hairdressers in the United States? How much time does a person need to allow for an appointment at the barber or hairdresser?

19. What are the special privileges of age and/or sex?

20. If people are invited to dinner, should they arrive early? On time? Late? If late, how late?

21. On what occasions would a person present (or accept) gifts from people in the country? What kind of gifts would they exchange?

22. Do some flowers have a particular significance?

23. How do people greet one another (shake hands, embrace, or kiss)? How do they leave one another? What does any variation from the usual greeting or leave-taking signify?

24. If people were invited to a cocktail party, would they expect to find among the guests: Military people from the local forces? Officers, NCOs, or both? Foreign business people? Men only? Men and women? Local business people? Local politicians? National politicians? Politicians' spouses? Teachers or professors? Bankers? Doctors? Lawyers? Intellectuals, such as writers, composers, poets, philosophers, religious clerics? Members of the host's family? (Including in-laws?) Movie stars? Ambassadors or consular officials from other countries?

25. What are the important holidays? How is each observed?

26. What are the favorite leisure and recreational activities of adults? Teenagers?

27. What sports are popular?

28. Is television an important influence? What kinds of television programs are shown? What social purposes do they serve?

29. What is the normal work schedule? How does it accommodate environmental or other conditions?

30. How does the advisor's financial position and living conditions compare with those of the majority of people living in this country?

31. What games do children play? Where do children congregate?

32. How are children disciplined at home?

Figure B-1. Cultural analysis questionnaire (continued)

33. Are children usually present at social occasions? At ceremonial occasions? If they are not present, how are they cared for in the absence of their parents?

34. At what age are children considered adults? How does this society observe a child's "coming of age?"

35. What kind of local public transportation is available? Do all classes of people use it?

36. Who has the right of way in traffic (vehicles, animals, or pedestrians)?

37. Is military training compulsory?

38. Are the largest circulation newspapers generally friendly in their attitude toward the United States? Radio and TV broadcasters?

39. What is the history of the relationships between this country and the United States?

40. How many people have emigrated from this country to the United States? Other countries? Are many doing so at present?

41. Are there many American expatriates living in this country?

42. What kinds of options do foreigners have in choosing a place to live?

43. What kinds of health services are available? Where are they located?

44. What are the common home remedies for minor ailments? Where can medicines be purchased?

45. Is education free? Compulsory?

46. In schools, are children segregated by race? By caste? By class? By sex?

47. What kinds of schools are considered best (public, private, or parochial)?

48. In schools, is learning conducted by memorization and repetition or by reasoning and understanding?

49. How are children disciplined in school?

50. Where are the important universities of the country? If university education is sought abroad, to what countries and universities do students go?

Figure B-1. Cultural analysis questionnaire (continued)

COUNTERINSURGENCY

B-2. Advisors conducting COIN operations must thoroughly research the AO, the counterpart, and the insurgent force. Figure B-2, page B-4, provides a number of questions designed to spur the research process. As with Figure B-1, this figure is not all-inclusive, but is as a starting point for further study.

1.   What is the insurgent structure of government in the area, the politico-guerrilla apparatus? Does the advisor have this on an overlay of a map of his area, showing traditional local boundaries upon which the enemy usually bases his structure? What is the status of the insurgent village government? Are there any recent changes (purges) in his local leadership?

2.   Are the advisor's plans against this insurgent threat truly realistic? Is the advisor planning things that not only can be done, but also stay done? What is needed to have them stay done? Is this included in the plans?

3.   How does the advisor see the counterpart: as a man, or simply as an official? Does the advisor deal with him as one official to another, as friends bound by adherence to a common goal, or a combination of the two? Does the advisor do the counterpart's work for him and make his leadership weaker, or carefully influence him toward making his organization really produce?

4.   How does the counterpart stack up in the minds of the people of the area when compared with his insurgent opponent? Because the people are the target of the campaign, what does the advisor's answer to this question suggest that they do next?

5.   Does the advisor know whether his counterpart is reasonably honest or is corrupt? If the advisor does not know, how can he find out skillfully? If the counterpart is corrupt, is there something the advisor can do constructively (other than reporting to HQ) to shape him up?

6.   Does the advisor discuss principles of good service to the nation with his counterpart, so that the advisor and counterpart are aware of the moral strengths required in a campaign against an enemy whose officials may claim to serve with disciplined honesty? If the enemy is characterized by corrupt practices, does the advisor make his contempt known to his counterpart?

7.   Do the advisor and his counterpart just talk with other officials or do they talk with the nonofficial people in the AO to find out their true feelings?

8.   Do the friendly forces in the AO practice courtesy among the people in both urban and rural settings, at rest stops, and along the roads? Does this go along with the advisor's own ideas of military courtesy? Are there ways the advisor can help improve it?

9.   Are the pay, rations, and allowances of local civil and military personnel enough to live on or do they have to resort to theft to get by? If something needs changing, what is it?

10. Does the advisor's plan make use of all resources in the AO? Does the advisor know all the local government leaders in the AO? Who are the local leaders outside the government whom the people respect, the business leaders, and the village or municipality leaders? Are they part of the advisor's plan to obtain success?

11. Does the advisor's planning strengthen and encourage the growth of individual self-reliance? Do the people in the towns, hamlets, or villages ever hold town meetings to decide their own affairs, with a real opportunity to speak frankly and without fear of reprisals? Do individuals get a chance to work for their own independence? Do farmers have a way to make more money from crops; get credit without outrageous repayment rates; get a fair deal on seeds, fertilizer, and equipment; and market what they produce?

Figure B-2. Counterinsurgency questionnaire

RECOMMENDED PRACTICES IN DEALING WITH COUNTERPARTS

B-3. The SF advisor may adopt a number of practices to help in dealing with counterparts. Advisors should—

- Assure that the SF presence is understood.
- Find a basis for common interest with the local people.
- Try to understand why things are done the way they are done.
- Start with where the people are and what the people want.
- Work within the local cultural framework.
- Help people believe they can improve their situation.
- Be content with small beginnings.
- Utilize local organizations and recognize their leaders.
- Help the government get organized to serve the people.
- Train and utilize subprofessional, multipurpose local workers.
- Expect slow progress.
- Transfer controls constructively.
- Expect no gratitude from those helped.

ASSURE THAT THE SPECIAL FORCES PRESENCE IS UNDERSTOOD

B-4. An advisor entering an AO works under the sponsorship of a military commander, mayor, village head, or some other recognized local leader. Prior to this, the advisor's arrival may need to be coordinated through the district, sector, or other appropriate local office. Clearances from the distant national, state, province, or sector governments cannot compensate for a clear, local explanation of why an American Special Forces Soldier is in the area. This is especially true of small, isolated communities where it is unusual for a stranger to appear for even an hour without being acknowledged and accepted by local leaders. Without explanations from locally respected persons, the local population will arrive at its own explanations, often to the detriment of the SF effort.

FIND A BASIS FOR COMMON INTEREST WITH THE LOCAL PEOPLE

B-5. If the advisor shows appreciation for the local people as individuals, culture gaps and language barriers can be overcome and common ground can usually be found. An advisor should listen when the locals speak and should show interest in the things that they show him. Initial conversations usually center on universal matters, such as food, shelter, clothing, health, and education. Over time, discussions may be naturally brought around to the matter the advisor wants them to consider. An advisor will be better received if he knows something about earlier HN contributions in such matters as agriculture, folk art, religion, and architecture. Naturally, the advisor will be more effective and appreciated if he can speak the local language.

TRY TO UNDERSTAND WHY THINGS ARE DONE THE WAY THEY ARE DONE

B-6. Although some local practices may seem strange at first, they generally have good reasons behind them. Advisors can discover these reasons with careful observation and a creative imagination. Food habits, family traditions, folk cures, and festive celebrations almost always have a great deal of human experience at their root. The advisor also needs to be aware that many villages contain rival subgroups and factions. This tension needs to be accounted for when working with the local people.

START WITH WHERE THE PEOPLE ARE AND WHAT THE PEOPLE WANT

B-7. The lives of traditional people across the world are usually simple and realistic. It is important to find out what the local people really want most, and to work with the people to achieve this aim. The local people may want a public school or a road, whereas the SF advisor thinks the village most needs a well or a clinic. The need that the local people feel often is the best starting point, regardless of its comparative merits. Once this aim is addressed, people are more likely to be appreciative and cooperative—they begin to raise their expectations and become interested in working for other improvements. In order to address the initial desires of the people, the advisor may need to call in personnel with the specialized skills needed for the particular project. Although this can result in somewhat of a delay, it will help the advisor to achieve greater cooperation over time.

WORK WITHIN THE LOCAL CULTURAL FRAMEWORK

B-8. The SF advisor needs to understand such basic cultural matters as the ethnic backgrounds of the people, family relationships, leadership patterns, value systems, and the technological level of the people (as related to ways of making a living). He also needs some knowledge of local services such as health, education, communications, and transportation. Many things will depend upon the advisor's realism in the cultural field (for instance, the extent to which locally available physical resources can be used).

HELP PEOPLE BELIEVE THEY CAN IMPROVE THEIR SITUATION

B-9. The vast majority of the traditional peoples of South America, Asia, and Africa live in a largely static environment. Through their experience with change, they are more fearful of losing status than they are hopeful of bettering their condition. Therefore, changes suggested by advisors are often viewed with fear. Concrete local projects that yield easily observed benefits are helpful in convincing the villagers that they can improve their situation and make them more willing to cooperate in other projects.

BE CONTENT WITH SMALL BEGINNINGS

B-10. Change tends to come slowly in areas where there have been few in recent times. It is good to remember that, historically speaking, most scientific developments in the West occurred only recently. The advisor should keep in mind that knowledge—technical or otherwise—is cumulative. Once a small beginning has been made, greater activity and additional changes soon follow. It is easier to achieve momentum than it is to maintain it. The important thing is for the advisor to make a start within as promising a framework as possible and with the support needed to sustain the momentum achieved.

UTILIZE LOCAL ORGANIZATIONS AND RECOGNIZE THEIR LEADERS

B-11. People respond best when their local organizations are recognized as important and useful. A program is unlikely to succeed unless it is carried forward within the local organizational framework. The recognized local leaders must be consulted and encouraged to make contributions as they can. A well-conceived technical activity reflects credit on the local leaders associated with it. Attention must be given not only to the officials and the family heads of local groups, but also to the quiet, behind-the-scenes leaders. The surest way for an activity to be continued after the SF advisor leaves is for it to have been launched and carried forward within the local organizational and leadership framework.

HELP THE GOVERNMENT GET ORGANIZED TO SERVE THE PEOPLE

B-12. In order for the advisor to be most effective, he must understand the local government organization and how his activity fits into the overall scheme. There should already be a set of agreements between various local agencies and the national government (usually through some sort of interministerial council) that provides for a coordinated effort in servicing the varied needs of the local people. The SF advisor may need to work with appropriate agencies to assist in getting such agreements made. If such agreements already exist, SF advisors should be careful to recognize and strengthen them. The work of the advisor in one field is most meaningful when properly coordinated with the contributions of individuals and agencies in other fields.

TRAIN AND UTILIZE SUBPROFESSIONAL, MULTIPURPOSE LOCAL WORKERS

B-13. Selected young people in the villages can be trained and used as subprofessional, multipurpose village workers to enable the advisor to make the best use of his time. Otherwise, the SF soldier's influence is restricted to where he is standing and the immediate vicinity. Furthermore, the advisor may spend so much of his time establishing and maintaining enough rapport with the villagers, he may become incapable of rendering any real service at all. The gap between the local population and the SF advisor is usually a formidable one because of the great educational and cultural differences between them. Often the advisor works with villagers who are poor, illiterate, and devoid of outside contacts. Volunteer or paid local workers (who serve as liaison between the villagers and the advisors) have proven of great help in getting the benefits of subject-matter technical activities.

EXPECT SLOW PROGRESS

B-14. As the local people begin to see successes from their joint efforts and begin to have new hope, they naturally want a larger hand in their own matters. The advisor may sometimes feel they want to assume more responsibility than they are able to carry. These evidences of growing pains should be appreciated, for they are a necessary part of becoming able to assume responsibility. They indicate that the local population is beginning to believe they can do more and more things for themselves. The advisor needs to adjust to the growing desires of the people to help themselves.

TRANSFER CONTROLS CONSTRUCTIVELY

B-15. The matter of institution-building is a challenge to SF advisors. They need to help the local people see how they can build the new—what they want—upon the foundations of the old—what they already have. From the beginning of a project the team needs to envision, at least roughly, and discuss with local leaders the various types of training of local personnel needed, the means by which needed financial support can be had, and the several progressive transfers of responsibilities that are to be made before the full operation of the activity can be relinquished. If operating responsibility is transferred too early there likely will be some breakage—usually of material things. If, on the other hand, the team keeps control too long, the local people who have wanted to take over may become disillusioned with them (or even hate them) for not relinquishing control to them when they thought it should have been. The team must work out with local leaders the timing of the phasing out each technical activity that is started.

EXPECT NO GRATITUDE FROM THOSE HELPED

B-16. People who benefit from assistance sometimes feel defensive. In accepting assistance they are, in a sense, admitting their own insufficiency. The self-esteem of a person, community, or a nation as a whole is a delicate thing. The team should not, therefore, expect thanks. Rather, the team should approach the people in a spirit of fraternity and humility, taking satisfaction in the progress they may make. The team should do its job the best it can and accept work well done as is its own reward.

This page intentionally left blank.

Appendix C

Working With Translators and Interpreters

SELECTION AND USE OF INTERPRETERS

C-1. The use of interpreters is an unsatisfactory substitute for direct communication, but there are situations where they may be necessary. These individuals may be the only link between the advisor and his counterpart (or other important local nationals). The SF advisor must strive to maximize interpreters' strengths and anticipate their weaknesses.

AGE, GENDER, AND RACE

C-2. Factors of gender, age, and race, are potentially troublesome and can seriously affect your mission. Because differences from country to country vary greatly, advisors should check with the in-country briefing teams for specific taboos or favorable characteristics. In certain cultures, for example, the status of females in the society is such that they should not be used as interpreters with male sources.

TIME

C-3. When planning communications, advisors must allow additional time for interpreters to do their job. A 10-minute conversation may take up to 30 minutes, depending on the interpreter's ability. Advisors should plan accordingly when scheduling meetings.

RANK AND MILITARY AFFILIATION

C-4. Advisors must consider rank when planning for interviews. If the interviewees are officers, it would be better to have an officer or civilian act as interpreter. If the sources are enlisted, an officer interpreter might intimidate them and stifle participation and interaction.

C-5. Interpreters are usually contracted civilians. They may not be accustomed to military methods, hardships, discipline, and courtesies.

OTHER CONSIDERATIONS

C-6. When working with interpreters, there are a number of other factors that the advisor must consider. Advisors should—

- Try to use two-man interpreter teams, as one may catch what the other missed or forgot.
- Avoid organizing interpreters into interpreter pools. This detracts from their ability to react quickly to unexpected situations.
- Keep OPSEC in mind and assume the interpreter's first loyalty is to his country, not the United States.
- Prepare the interpreter for technical terms. The interpreter must know the subject area and translate the advisor's meaning as well as his words.
- Establish rapport with the interpreter through personal contact. Without a cooperative, supportive interpreter, the mission could be in serious jeopardy.
- Learn the interpreter's background and show genuine concern for his family, aspirations, and education.
- Double-check that the interpreters understands. Many may attempt to save face by purposely concealing their lack of understanding.

- Instruct the interpreter to mirror the tone and personality of advisor's speech.
- Instruct the interpreter not to interject his own questions or personality.
- Avoid such phrases as, "Tell him that..." and "I would like to have you say..."
- Avoid looking at the interpreter during discussion; remain focused on the other person.
- Break thoughts into small, logical, translatable segments.
- Avoid idioms, jokes, military jargon, and slang.
- Control the interpreter. The advisor should inform him or her to never ask questions of their own and to never paraphrase the interviewer's question or the source's answers.
- Instruct the interpreter to never hold back information given by the source. Such withholding, regardless of how insignificant the interpreter may feel it to be, may adversely affect the conversation and the mission.
- Test the interpreter periodically for accuracy, loyalty, and honesty. Control of the interpreter is increased if he knows that he is periodically tested.
- Avoid bullying, criticizing, or admonishing the interpreter in the presence of the source. The advisor should criticize in private to avoid lowering interpreter's prestige, and thereby impairing his effectiveness.

Checking Interpreters

We were checking on the condition of a group of elderly people in a Croat village and I was in the background. The interpreter, who didn't think I could hear, asked the people if they'd had any problems or had they been harassed recently. They said they had, that a group of soldiers from a local unit had been around stealing and threatening to kill them if they reported anything. And then the interpreter said to the team leader, 'No, these people have had no problems.

You have to realize you're completely at the mercy of the interpreters if you don't possess some understanding of the language—at the mercy of people who may be discussing arresting you or taking you hostage, and you don't even realize that you are in danger.

Former SF Advisor, Bosnia

C-7. The use of interpreters and translators seems very straightforward—the advisor simply speaks in English and the interpreter repeats it in the appropriate language. However, translation should not be taken lightly. Obviously, language is vital to communication across cultures, but language can also be the source of many misunderstandings in intercultural communication.

C-8. Translators and (especially) interpreters are seldom native speakers of both languages. Normally, they are native speakers of one language with training in the other. This means that they have a good command of the formal language when used correctly and by a speaker with an accent they are familiar with.

CHALLENGES WHEN WORKING WITH INTERPRETERS

C-9. There are a number of challenges the advisor must overcome when working with interpreters. These challenges include—

- Translation.
- Pronunciation.
- Word choice and meaning.
- Slang and idiomatic expressions.

TRANSLATION

C-10. In communicating with people of a different culture, advisors should use at least a few words or phrases in their language. It is even better to have sufficient command of the language to be able to ensure that the interpreter or translator is not straying from the desired message.

C-11. There is a strong tendency to translate messages literally (word-for-word) from the speaker's language to the target language. This technique may work well, but it can lead to embarrassing situations.

PRONUNCIATION

C-12. Misunderstandings in intercultural communication can arise even when two cultures use the same language. Even though English is used widely throughout the world, regional differences exist in pronunciation. This can make it difficult for speakers of English from two different countries—or even two different regions of the same country—to understand one another.

C-13. English is one of the official languages of Singapore. It is often used as a lingua franca—for official communication—between the various language subgroups on the island and is often used by government officials and businessmen to communicate with their countrymen. However, English-speaking foreigners frequently have difficulty understanding because of the differences in pronunciation. To Americans, Singaporeans do not clearly pronounce words with similar beginning and ending sounds. Words such as *tree* and *three*, or *pen* and *pan*, become garbled. This can lead to misunderstandings despite sharing a common language.

WORD CHOICE AND MEANING

C-14. Advisors must use caution in selecting words to translate. Ambiguous and unfamiliar words should be avoided.

Ambiguous Words

C-15. The same word may have different interpretations in different cultures. In most parts of Asia, for example, the word *family* refers to parents, siblings, grandparents, uncles, aunts, cousins, and so on. To an American or European, *family* usually refers to the immediate family. If two colleagues—one Asian and one American—were to carry on a conversation about their families, they may think that they are talking about the same thing, but actually they are not.

Unfamiliar Words

C-16. The use of unfamiliar words can sometimes lead to expensive mistakes. For example, in a fictional discussion between an American businessman and a Japanese customer, the American concludes by saying, "Well then, our thinking is in parallel." They bid goodbye, but months pass without further word. Finally, the frustrated American inquires as to why. The Japanese customer replied, "You used a word I did not understand—*parallel*. I looked it up in my dictionary and it said parallel means two lines that never touch." The customer concluded that the American businessman thought their positions were irreconcilable.

C-17. In order to avoid miscommunications, advisors should take simple measures to achieve greater clarity in meaning. They choose words carefully, ensuring they are unambiguous and easily understood. Qualifications and definitions should be provided for terms that are likely to be misunderstood. Finally, advisors should never assume that the message was correctly understood. Instead, they should ask for feedback to ensure the audience clearly understood the message as intended.

SLANG AND IDIOMATIC EXPRESSIONS

C-18. Cultures develop their own slang and idiomatic expressions that may be foreign to other cultures using the same language. For example, Australians commonly refer to a friend as a "mate," but in the United States, the word "mate" primarily refers to a spouse. Idiomatic expressions can be especially confusing for non-native speakers who are not very proficient with the language. For example, "pulling someone's leg" is a common idiomatic expression for joking around, but an interpreter unfamiliar with this expression would be puzzled because, obviously, he had not even touched the speaker—let alone pulled on their leg. Because the world abounds in languages and dialects, and because it is often impossible to predict well in advance where a SF advisor may be sent, he may lack the linguistic ability to communicate effectively.

Appendix D

Nongovernmental and Intergovernmental Organizations

In the course of performing advisory duties, the SF Soldier frequently will work in environments alongside a large array of organizations, agencies, and groups. In addition to the more familiar joint and interagency environment of USG operations, the operational arena increasingly contains a wide variety of NGOs and IGOs.

MEANS, METHODS, ORGANIZATION, AND PHILOSOPHY

D-1. The means, methods, organization, and philosophy of most NGOs and IGOs are very different from U.S. military and governmental agencies. However, because these organizations play such a vital role in development and humanitarian assistance programs, the advisor must have some understanding of these differences.

D-2. NGOs are transnational organizations made up of private citizens that maintain a consultative status with the Economic and Social Council of the United Nations (UN). NGOs may be professional associations, foundations, multinational businesses, or simply groups of individuals linked by a common interest in humanitarian assistance activities. The term NGO has been expanded in recent years to include the full range of nongovernmental relief, development, and assistance organizations, regardless of UN status.

D-3. IGOs are organizations with global influence, such as the UN and the International Committee of the Red Cross (ICRC). These organizations normally have official status of some kind, to include defined privileges and responsibilities under international law.

D-4. When dealing with NGOs and IGOs, the SF advisor should recognize a number of characteristics that are common to most such organizations. For example, most NGOs and IGOs are just as committed to their cause as the SF Soldier is to his. These organizations usually stress impartiality, neutrality, and independence above all else. Members of NGOs and IGOs may have in-country experience that vastly exceeds that of the SF advisor, and they are essential in the transition to peace. These organizations are generally opposed to assuming a subordinate or junior-partner status with military personnel. NGOs and IGOs have no central command and vary widely in their willingness to work with the military. Some may be suspicious of the SF advisor and his purpose. Some may be willing to work near—but not with—the military. Some perceive information gathering as interrogation.

UNITED NATIONS AGENCIES AND OTHER INTERNATIONAL ORGANIZATIONS

D-5. The UN emergency-management apparatus, reorganized and streamlined in 1992, has humanitarian, development, political, and security components. On the humanitarian side, the United Nations Department of Humanitarian Affairs (UNDHA) is responsible for mobilizing and coordinating the collective efforts of the international community to meet human needs in disasters and emergencies and to facilitate the smooth transition from relief to development. Other UN humanitarian agencies include the United Nations High Commissioner for Refugees (UNHCR) and the World Food Program (WFP). These agencies respond to specific emergencies at the direction of the United Nations Security Council (UNSC) and member countries.

D-6. The United Nations International Children's Emergency Fund (UNICEF), the World Health Organization (WHO), and the United Nations Development Program (UNDP) are the UN's development organizations, dealing with long-term humanitarian issues, but generally not relief efforts. All of these programs work both with their own staffs and with individual NGOs that implement UN programs in the field. The United Nations Department of Political Affairs (UNDPA) follows political developments worldwide, so as to provide early warning of impending conflicts and analyze possibilities for preventive action by the UN. The United Nations Department of Peacekeeping Operations (UNDPKO) was greatly expanded in 1992 to include monitoring, planning, and support of operations. It also serves as the Secretary General's military staff. The UNDPKO is responsible for the military, civilian police, and electoral components of a complex mission.

D-7. Other public IGOs in this category include a number of regional government organizations, such as the Organization of African Unity (OAU), the Organization of American States (OAS). Also included are subregional groups, such as the Economic Community of West African States (ECOWAS), which is very active in Liberia. As with the UN agencies, these IGOs are characterized by their special status as legal entities under certain tenets of international law.

PRIVATE NONGOVERNMENTAL ORGANIZATIONS

D-8. Private international organizations are, in effect, groups of NGOs. They are usually composed of individual national chapters and include worldwide and regional institutions involved in humanitarian missions. Examples of private international organizations include the ICRC, the League of Red Cross, and Red Crescent Societies. These organizations operate around the world independently of any government, and may also enjoy special status as legal entities under international law.

D-9. Donor agencies are primarily national government funding organizations that provide official resources for development and relief. They also provide much of the funding for NGOs. The principal donor agencies represent national governments directly or indirectly and include USAID, the Canadian International Development Agency (CIDA), the Japan International Cooperation Agency (JICA), the U.K.'s Overseas Development Agency, and the European Community Humanitarian Organization (ECHO), which coordinates the efforts of several European government agencies. The World Bank and regional development banks are also donor agencies responding to the guidance of their multiple members. Although the banks do not play a role in relief efforts, they are increasingly seeking ways to be responsive during reconstruction.

D-10. There are thousands of individual NGOs operating around the world. Some, like the International Rescue Committee (IRC), World Vision, Cooperative for Assistance and Relief Everywhere (CARE), Christian Children's Fund, Save the Children, and Catholic Relief Services, are registered in the United States and conduct their missions overseas. Others, like Oxford Famine Relief (OXFAM) and Médecins sans Frontières (Doctors Without Borders) (MSF), operate out of other developed countries and partake in activities around the world. Other NGOs are indigenous to the countries where relief and development needs exist.

D-11. NGOs differ in size, resource base, thematic and geographic focus, and access to and use of technology, among other things. Some NGOs are quite large—for example, CARE's total support and revenue top $450 million—whereas many others have operating budgets of less than $10,000. The origin of funding may also vary greatly from NGO to NGO. These organizations may be funded by—

- Public resources, such as grants and contracts from donor government agencies and international organizations.
- Private resources, such as contributions from individuals, religious groups, communities, foundations, and businesses, in the form of either money or gifts-in-kind.

D-12. Most NGOs rely upon a combination of public and private funding; however, some NGOs decline to accept government funding so as not to be compromised by specific government policy interests. NGO personnel rosters vary according to funding and mission. Larger NGOs have a greater ability to respond to unexpected contingencies because of their increased resource base.

D-13. A 1995 UN report on global governance suggested that there were approximately 29,000 international NGOs. Domestic NGOs have grown even faster. In Russia, where almost none existed before the fall of communism, there are at least 65,000 such organizations. Dozens more are created daily around the world.

D-14. Some NGOs are primarily helpers—distributing relief where it is needed. Others NGOs are mainly campaigners, existing to promote issues their members deem important. The general public tends to see them as uniformly altruistic, idealistic, and independent. This is not always the case.

D-15. A growing share of development spending, emergency relief, and other aid passes through NGOs. USAID refers to NGOs as "the most important constituency for the activities of development aid agencies." Much of the food delivered by the WFP, a UN body, is actually distributed by NGOs. Between 1990 and 1994, the proportion of European Union (EU) relief aid channeled through NGOs rose from 47 to 67 percent. The ICRC estimates that NGOs now disburse more money than the World Bank.

THE BIG EIGHT

D-16. Eight major federations of international NGOs control a majority of worldwide annual relief funding. Known collectively as "the big eight," these organizations (in no particular order) are—

- CARE.
- World Vision International.
- OXFAM.
- MSF.
- Save the Children Federation.
- Eurostep.
- Coopération Internationale pour le Développement et la Solidarité (International Cooperation for Development and Solidarity or CIDSE).
- Association of Protestant Development Organizations in Europe (APDOVE).

TYPES OF NONGOVERNMENTAL ORGANIZATIONS

D-17. By their very nature, most NGOs are charitable. Some NGOs involve a top-down effort that demands little participation by the recipients. This type of NGO conducts activities directed toward meeting the needs of the poor, for example, the distribution of food, clothing, or medicine, or the provision of housing, transport, and schools. NGOs of this type may also undertake relief activities during natural or man-made disasters. Such NGOs often are criticized as being overly paternalistic—promoting dependence, rather than local development and autonomy.

D-18. Other NGOs are participatory. These organizations are characterized by self-help projects where local people are involved, particularly in the implementation of a project by contributing cash, tools, land, materials, and labor. In the classical community-development project, participation begins with the need definition and continues into the planning and implementation stages. Cooperatives often have a participatory orientation.

D-19. The aim of such participatory NGOs is to help poor people develop a clearer understanding of the social, political, and economic factors affecting their lives, and to strengthen their awareness of their own potential power to control their destiny. Sometimes, these groups develop spontaneously around a problem or issue. At other times, outside workers from NGOs facilitate their development. Regardless of their conception, this type of organization is characterized by maximum involvement of the people, with the NGO acting as a facilitator.

D-20. Many community-based organizations emerge out of individual and collective initiative. These might include sports clubs, women's organizations, neighborhood associations, religious groups, or educational organizations. There are a large variety of these organizations—some are supported by NGOs, IGOs, or bilateral or international agencies; others operate independently. Some community-based organizations are devoted to raising the consciousness of the urban poor or helping them to understand their rights in gaining access to needed services; others are involved in providing such services.

ADVANTAGES OF NONGOVERNMENTAL ORGANIZATIONS

D-21. NGOs may enjoy a number of advantages over their military counterparts when operating in a foreign region. Some of these advantages include the following:

- They have the ability to experiment freely with innovative approaches and, if necessary, to take risks.
- They are flexible in adapting to local situations, responding to local needs, and developing integrated projects.
- They enjoy good rapport with people and can render micro-assistance because they can identify those who are most in need and tailor assistance accordingly.
- They have the ability to communicate at all levels, from the neighborhood to the top levels of government.
- They are able to recruit both experts and highly motivated staff with fewer restrictions than the military or OGAs.

DISADVANTAGES OF NONGOVERNMENTAL ORGANIZATIONS

D-22. Along with the advantages NGOs enjoy, there are a number of disadvantages that they must overcome. These disadvantages include the following:

- Paternalistic attitudes that restrict the degree of participation in program or project design.
- Restricted or constrained ways of approach to a problem or area.
- Reduced replicability of ideas and programs due to uniqueness or narrow focus of the project or selected area, relatively small project coverage, and dependence on outside financial resources.
- Territorial possessiveness of an area or project between agencies; such "turf wars" are common.

MILITARY-NONGOVERNMENTAL ORGANIZATION INTERACTION

D-23. NGOs cannot be viewed as a single, homogeneous group. They represent many countries and many organizations throughout the world, each with its own agenda and desired outcome of assistance. Some NGOs are very cooperative and work well with the military. Others, such as the ICRC, accept military help with great reluctance. Others are unwilling to cooperate with the military at all, such as MSF. MSF is very active in humanitarian assistance and disaster relief work worldwide; however, it has stated that NGOs cannot "maintain their impartiality and independence of action once they agree to act in coordination with, or effectively under the coordination of, a military structure." Some NGOs may continue to pay bribes and extortion fees to bandits, even while military forces are in the area to protect them. These organizations consider the long-term scenario; they believe that if they don't pay the fees now, things will be worse after the military forces redeploy.

D-24. Military forces cannot exercise C2 over NGOs; however, a spirit of cooperation must exist between the two in order to be successful. In some ways, the military assumes the unusual role of service provider. Military forces often support and secure NGO operations. Humanitarian agencies do not (or should not) want the military to do their jobs for them—they simply need protection and an opportunity to do their jobs. The establishment of civil-military operations centers (CMOCs) is a key step in providing this support.

Planning Operations With Nongovernmental Organizations

D-25. Operation UPHOLD DEMOCRACY (Haiti, 1994–1995) marked the first time the USG organized to develop an interagency, politico-military plan of operations prior to undertaking a crisis response. The Office of U.S. Foreign Disaster Assistance (OFDA) and, through it, the NGO community, were brought into the USG and military planning processes at a relatively early stage.

D-26. NGOs often complain that USG or UN objectives are unclear, thereby hampering planning efforts. Because of domestic political considerations, member states sometimes prefer that UN objectives be purposely vague. This lack of clarity, however, complicates the planning process for all involved, including NGOs. NGOs are concerned about planning for participation in humanitarian-relief aspects of the emergency response, and not with military planning. These organizations complain that once the military is involved, the potential for disrupting informal—albeit effective—channels of communication among NGOs is considerable. Many NGOs do not favor the military's standardized approach to problems, believing that it diminishes the flexibility that NGOs enjoy.

D-27. Given the size and independent nature of the NGO community, there are too many moving parts for a consistent, closely coordinated effort. Once numerous actors become involved, coordinating operations becomes very complicated (for example, there were over 400 independent NGOs in Haiti during ENDURING FREEDOM). Commander's intent is critical to military planning. Many NGOs cannot identify who the commander is, let alone the commander's intent.

D-28. In Operation JOINT ENDEAVOR (Bosnia 1995–1996), North Atlantic Treaty Organization (NATO) programs sought to identify normality indicators as a guidepost to assess the effectiveness of NATO initiatives. Most NGOs believed such indicators to be irrelevant and even misleading. Because local procurement contracts tended to inflate normality indicators, the very presence of the military skewed its own indicators and masked real development problems. The NGO presence also affected local normality indicators, although perhaps on a different scale.

Coordination

D-29. Successful resolution of a crisis requires collaboration across the politico-military spectrum. Both the military and NGO communities have discovered that the establishment of CMOCs is helpful in furthering operational-level coordination and communication.

D-30. Many NGOs do not understand—and may even be suspicious of—the military's seemingly endless demands for detailed information on NGO operations. One unidentified NGO field director stated, "Even my headquarters in New York doesn't demand the kind detailed reporting the military does." Some NGOs are openly skeptical about the way the military might use the provided information.

Training

D-31. Several training initiatives have helped foster better relations between NGOs, USG agencies, and the U.S. military. All organizations must deal with complications created by personnel changes and the need to keep training requirements current. The U.S. Army's Joint Readiness Training Center (JRTC) at Ft. Polk, Louisiana, conducts training exercises across the range of military operations and incorporates information about working with the NGO community into some of its exercises. JRTC has begun to include members of the NGO community in its field exercises; however, must NGOs have limited budgets and cannot afford to send key personnel to extended military training exercises.

Perspectives

D-32. The vastly different objectives and perspectives that the NGO, USG, and military communities bring to a crisis can be both beneficial and troublesome. Typically, the NGO community has been involved in relief and development activities in a given country long before a crisis develops. The evolving crisis prevents the NGOs from carrying out their mission. As the situation deteriorates and gains international attention, the military begins to plan a response. When political authority determines to deploy military force to stabilize and provide security for humanitarian operations, it usually provides the armed forces with a limited mission and, often, a limited timeframe in which to complete the mission. Once in the field, the military mission (to provide security to the overall operation) and the NGO mission (to carry out specific relief activities and return to normalcy) may conflict. Although both communities have a common goal in their response to the complex humanitarian emergency (stabilization of the situation and a return to normalcy), each has a different perspective on how the goal should be achieved and how long it should take. NGO expectations regarding the military's role in providing security for their operations may not be realistic if they do not fully understand or appreciate the military's mission, doctrine, or approach to the use of force.

Communication

D-33. Effective communications are challenging during any operation involving NGOs and military forces. There are situations, for example, in which an NGO will be reluctant to communicate its plans to the military. NGOs express concern that in the process of sharing their plans with the military, they would be broadcasting their intentions to indigenous armed groups.

D-34. Communications that are too integrated may also affect the HN perspective on the relationship between an NGO and the military. Host governments may become suspicious when there is too much interoperability among IGOs, NGOs, and the military. Many NGOs are reluctant to establish intricate communications nets in order to protect their neutrality. Additionally, in contrast to the military's high-tech approach, much of the NGO community uses antiquated equipment manned by locals.

D-35. NGOs and IGOs tend to cause the most trouble for the military at checkpoints; some may even try to run through checkpoints. Checkpoints should be designed to prevent this. One common irritation for NGOs is redundant checkpoints that require them to undergo numerous checks of their vehicle and belongings. NGOs should not be given preferential treatment at checkpoints; however, SF Soldiers must exercise restraint in the use of force. ROE help to identify the appropriate response. Most point out that lethal force should not be employed unless the Soldier is danger of injury or death.

D-36. Most NGO field staffs have established local connections. In an emergency, on-site personnel doing development work may quickly shift their resources to relief work. The relationships established by NGOs can be immensely helpful to newcomers. NGOs operating in the area are likely to have—

- Functioning offices.
- Communications facilities.
- An indigenous staff and local contacts, which can help newcomers understand—
- The local power structure.
- The shifting allegiances of particular political players.
- The hazards that relief workers are likely to encounter.
- The potential blunders relief workers are likely to make.

D-37. Because of their longstanding relationships, NGOs may also be able to provide valuable information about the details of particular area or culture and the relative dependability of various local groups and individuals. NGO knowledge of a situation and of a locale can often yield intelligence; however, NGOs may be very sensitive about sharing information if it is collected under that label. They are likely to see their long-term success as dependent upon good will and open relationships with the indigenous population. This can make them wary of compromising any trust they have established by telling too much to the military or even being seen too often with military personnel. It is essential that NGO and military personnel—each of whom is likely to have information useful to the other—create a way of sharing information that addresses the inhibitions they each may have about disclosure.

Note: Additional information about NGO and IGO structure, capabilities, and coordination may be found in JP 3-08.

NONGOVERNMENTAL ORGANIZATION CULTURE

D-38. In order to operate—and cooperate—effectively, SF advisors must have an understanding of NGO culture, their peculiarities, and their attitudes toward U.S. and foreign government and military agencies.

Neutrality and Advocacy

D-39. The ideas of neutrality and advocacy are deeply rooted parts of the NGO culture. Most relief operations follow a strict policy of political neutrality. They also consider themselves to be the advocates of their clients. The majority of NGOs, including those based on a religion, believe their purpose is to relieve human suffering regardless of political, ethnic, religious, or other affiliation. This policy is exemplified by most of the development and relief NGOs and by NGOs that focus specifically on conflict-resolution work. The latter believe that being able to establish a common ground between antagonists (who can then resolve their own dispute) relies upon their being perceived as having no preference for either side of a conflict.

D-40. When a conflict is well defined and opposing militaries are fighting each other, there exists some clarity as to what constitutes neutral behavior. When civil conflict includes the targeting of civilian populations, or when a strategy includes the denial of food to a region (even if it does not include a physical attack), NGO efforts to provide relief will not be seen as neutral. NGOs can even find themselves bringing aid and support to groups who are in desperate need, but whose politics are well known and abhorrent to the international community. In both Somalia and Rwanda, relief organizations found themselves in a situation where their relief work among perpetrators of a conflict likely prolonged the conflict.

D-41. Some NGOs have faced the neutrality dilemma by acknowledging that, in some instances, neutrality may not be possible. To pretend that it is may assist oppressors. One such organization is the Mercy Corps. This NGO has declared, "Mercy Corps will not seek out occasions for advocacy, but will respond to those circumstances when advocacy is the only responsible act." Mercy Corps considers its clients the impoverished and the oppressed. Results, Inc., is another NGO that not only has abandoned neutrality, but actually embraces advocacy. Its aim is to alleviate poverty, and it seeks influence through press conferences, editorials, and other lobbying activities.

D-42. Most human rights NGOs are politically active and seek to change conditions in the countries in which they work. Although neutrality has never been part of their creed, human rights NGOs must be careful of image they portray. If they are seen as political, as choosing one side over the other, NGOs can become targets.

Characteristics

D-43. Responsiveness in crisis is a trademark of humanitarian NGOs, and many are poised to work effectively when rules are unenforceable and authority nonexistent. This does not mean that these organizations do not make every effort to apply their experience and best thinking to a crisis. Indeed, the first step in mounting an NGO operation is to have trained staff make an assessment of the situation.

D-44. After the assessment, relief workers go to the field to accomplish a wide variety of tasks. These may include renting office space, hiring staff, planning projects, locating domestic suppliers, importing goods (to include dealing with customs and tax officials), hosting visitors (to include potential donors), working with the media, fact finding, and reassessing the mission.

D-45. Field workers share many of the hardships of local inhabitants. These may include a lack of electricity, lack of hot water, exposure to disease and illness, and even occasional artillery shelling. At the same time, field workers enjoy a key privilege that the locals do not—the field workers can always leave. In meeting their own needs, NGOs may inadvertently consume the best of available space, workers, and even food. Additionally, NGO transportation and communication facilities may be better than those of the local government. This can cause tension. It can also make NGO a target for ordinary crime (that is, theft of vehicles or relief supplies).

D-46. Security issues for NGOs are magnified when relief efforts are conducted amid armed conflict. Such conflicts bring military and NGO personnel into close contact. Both groups must strive to understand each other and work together; however, difficulties are common. Whereas some NGOs may see the military as offering valuable protection, others may see military security precautions as excessively cautious. Because most NGOs wish to maintain a nonthreatening image, they may perceive a close association with the military as endangering the protection their vulnerability affords. Some NGO staff and some of the local population may even see military forces as repressive, rather than protective.

D-47. As with those who join the military, those who work for NGOs are not seeking riches. Some members simply want to devote their energy and skills to an organization that—by definition—does not make a profit. Others may be caught up by the inspiration and ideals of a particular NGO. Still others are committed to a particular field (such as human rights), and working for an NGO is just one of several ways they are able to fulfill their purpose.

D-48. Compared to most corporate or governmental organizations, NGOs offer employment that maximizes independence, mobility, flexibility, variety, international travel, and the challenge of working in other cultures. NGO staff members are attracted by the challenge of managing programs in developing countries, often times in several languages.

D-49. Although some staff personnel are motivated by a desire to live their ideals, actually working in the field very quickly tempers idealism with reality. Even if a field staff lives better than the people with whom they work, they are likely to have a great deal of responsibility and to experience significant hardship and, at times, real danger. This is especially true when the government of an area has collapsed and no one is able to offer routine security or other essential functions.

D-50. As with any organization, field workers often experience tension with their HQ. Field staffs believe they are the ones with their "ears to the ground." Every field worker has stories about preposterous directions received from their HQ, and also knows about failed projects based on theory, not practice.

D-51. Field staff are likely to have the authority to design or commit to specific projects—at least at the level of providing seed money. If a new project holds promise, the HQ staff begins designing a proposal and seeking funds to permit fuller implementation. New funds may come from the organization's own capital; often, though, proposals are offered to the UN, USAID, The World Bank, ECHO, or other agencies.

D-52. Local workers may need permission to work for a foreign NGO. This can create problems, especially if they leave government work or seek military service exemption. Some remain government employees while working for an NGO, and may explicitly or implicitly become informants for that government.

D-53. Staff coming from outside the area may live in a secure (usually small) compound. Even in a semi-stable situation where friendly forces are in place, NGOs can be caught in precarious situations, such as when responsibility for operations in a particular area shifts from one organization to another (for example, from the UN to NATO), or from one nation to another (for example, from the United States to Australia).

D-54. Most NGO personnel are well educated. Permanent staff members usually college degrees, and many hold graduate degrees. Some work in their field of expertise, although other professionals may work as administrators (for example, a doctor working medical supply distribution may not be engaged in medical care at all).

D-55. Staff for large NGOs must have honed administrative and entrepreneurial skills. Whether for profit or not for profit, a large operation requires decisive action in difficult situations, skilled financial management, and firm control of a geographically dispersed enterprise. In addition, NGO managers must be aggressive fundraisers, and they must be able to fuse the energy and commitment of volunteers and paid staff.

D-56. In order to remain focused on those being served—refugees and victims—some NGOs refuse all public funding. Others expand their resources by accepting government monies, but may be uneasy and reluctant recipients. These NGOs fear distorted priorities as a result of governmental resource allocation—priorities which may be based on economic or political considerations rather than solely on human need.

D-57. In recent years, some NGOs have become more willing to plan, coordinate, and even to participate in postmission evaluations with other NGOs, UN agencies, governments, and military forces. For example, the group World Concern lists 74 agencies as its "partners." OXFAM America collaborates with such local NGOs as the Coalition for People's Agrarian Reform (Philippines) and the Eritrean Relief Association. Helen Keller International partners with the Aravind Eye Hospital in India and others.

D-58. When numerous NGOs are called to render relief in an emergency situation they may create a loose coordinating system by designating (or accepting) one of them as the lead (or coordinating) NGO. In Sarajevo, Catholic Relief Services assumed this role; in Haiti, Rwanda, and Somalia, CARE presided. Sometimes a lead agency will be charged with responsibility for umbrella grants from the USG.

D-59. NGOs do not have a hierarchical structure or chain of command. Work is accomplished through persuasion, not as the result of an order. To a military professional, this style may appear inefficient, irrational, or unreasonable; however, an effective NGO is capable of enlisting hands, hearts, and minds.

This page intentionally left blank.

Glossary

18A	Special Forces officer
18B	Special Forces weapons sergeant
18C	Special Forces engineer sergeant
18D	Special Forces medical sergeant
18E	Special Forces communication sergeant
18F	Special Forces intelligence sergeant
18Z	Special Forces operations sergeant
180A	Special Forces warrant officer
AAR	after action review
AECA	Arms Export Control Act
AO	area of operations
AOR	area of responsibility
APDOVE	Association of Protestant Development Organizations in Europe
AR	Army regulation
ARNG	Army National Guard
ARNGUS	Army National Guard of the United States
ARSOF	Army special operations forces
Benelux	Belgium, the Netherlands, and Luxembourg
C2	command and control
CA	Civil Affairs
CARE	Cooperative for Assistance and Relief Everywhere
CCDR	combatant commander
CCIR	commander's critical information requirement
CI	counterintelligence
CIA	Central Intelligence Agency
CIDA	Canadian International Development Agency
CIDSE	Coopération Internationale pour le Développement et la Solidarité
CJCS	Chairman of the Joint Chiefs of Staff
CMO	civil-military operations
CMOC	civil-military operations center
CN	counternarcotic
COA	course of action
COIN	counterinsurgency
COM	chief of mission
COS	chief of station
DA Pam	Department of the Army pamphlet

DATT	defense attaché
DCM	deputy chief of mission
DEA	Drug Enforcement Administration
DOD	Department of Defense
DOS	Department of State
DSAA	Defense Security Assistance Agency
ECHO	European Community Humanitarian Organization
ECOWAS	Economic Community of West African States
EEI	essential element of information
EPW	enemy prisoner of war
ESF	economic support fund
ETSS	extended training service specialist
EU	European Union
EW	electronic warfare
f2f	face-to-face
FAA	Foreign Assistance Act
FARC	Fuerzas Armadas Revolucionarias de Colombia
FBI	Federal Bureau of Investigation
FDC	fire direction center
FID	foreign internal defense
FM	field manual
FP	force protection
FSO	fire support officer
G-2	Army or Marine Corps component intelligence staff officer (Army division or higher staff, Marine Corps brigade or higher staff)
GCC	geographic combatant commander
GCIII	Third Geneva Convention
GSO	general services officer
GW	guerrilla warfare
HCA	humanitarian and civic assistance
HN	host nation
HQ	headquarters
IAW	in accordance with
ICRC	International Committee of the Red Cross
IDAD	internal defense and development
IGO	intergovernmental organization
IMET	international military education and training
IO	information operations
IPB	intelligence preparation of the battlefield
IR	information requirement
IRC	International Rescue Committee

ISR	intelligence, surveillance, and reconnaissance
JCS	Joint Chiefs of Staff
JFACC	joint force air component commander
JICA	Japan International Cooperation Agency
JP	joint publication
JRTC	Joint Readiness Training Center
JTF	joint task force
KIA	killed in action
KLE	key leader engagement
LTC	lieutenant colonel
MAAG	military assistance advisory group
MAP	Military Assistance Program
MARSOC	United States Marine Corps Forces, Special Operations Command
MEDEVAC	medical evacuation
METT-TC	mission, enemy, terrain and weather, troops and support available, time available, and civil considerations
MI	military intelligence
MLE	military liaison element
MLO	military liaison office
MOS	military occupational specialty
MOU	memorandum of understanding
MP	military police
MSD	Mobile Security Division
MSF	Médecins sans Frontières
MTP	mission training plan
MTT	mobile training team
NATO	North Atlantic Treaty Organization
NCO	noncommissioned officer
NGO	nongovernmental organization
NSC	National Security Council
O&I	operations and intelligence
OAS	Organization of American States
OAU	Organization of African Unity
OCONUS	outside the continental United States
ODC	Office of Defense Cooperation
OFDA	Office of U.S. Foreign Disaster Assistance
OGA	other government agency
OPATT	operational planning and assistance training team
OPCON	operational control
OPLAN	operation plan
OPORD	operation order

OPSEC	operations security
OSD	Office of the Secretary of Defense
OXFAM	Oxford Famine Relief
PAO	public affairs office
PATT	planning assistance training team
PDSS	predeployment site survey
PIR	priority intelligence requirement
PKO	peacekeeping operation
POI	program of instruction
POTUS	President of the United States
PRC	populace and resources control
PSO	post security officer
PSYOP	Psychological Operations
ROE	rules of engagement
S-1	personnel staff officer
S-2	intelligence staff section
S-3	operations staff section
SA	situational awareness
SAO	security assistance office
SecDef	Secretary of Defense
SECSTATE	Secretary of State
SF	Special Forces
SFG(A)	Special Forces group (airborne)
SFLE	Special Forces liaison element
SFODA	Special Forces operational detachment A
SFODB	Special Forces operational detachment B
SITREP	situation report
SME	subject-matter expert
SO	special operations
SOC	special operations command
SOCCE	special operations command and control element
SODARS	special operations debrief and retrieval system
SOF	special operations forces
SOFA	status-of-forces agreement
SOP	standing operating procedure
TACON	tactical control
TAFT	technical assistance field team
TAT	tactical analysis team
TC	training circular
TOC	tactical operations center
TTP	tactics, techniques, and procedures

UCMJ	Uniform Code of Military Justice
U.K.	United Kingdom
UN	United Nations
UNDHA	United Nations Department of Humanitarian Affairs
UNDP	United Nations Development Program
UNDPA	United Nations Department of Political Affairs
UNDPKO	United Nations Department of Peacekeeping Operations
UNHCR	United Nations High Commissioner for Refugees
UNICEF	United Nations International Children's Emergency Fund
UNSC	United Nations Security Council
U.S.	United States
USAF	United States Air Force
USAID	United States Agency for International Development
USAJFKSWCS	United States Army John F. Kennedy Special Warfare Center and School
USAR	United States Army Reserve
USASATMO	United States Army Security and Training Management Organization
USFK	United States Forces, Korea
USG	United States Government
USMC	United States Marine Corps
USN	United States Navy
USPACOM	United States Pacific Command
USSOCOM	United States Special Operations Command
USSOUTHCOM	United States Southern Command
USSR	Union of Soviet Socialist Republics
UW	unconventional warfare
WFP	World Food Program
WHO	World Health Organization
WO	warrant officer
WWI	World War I
WWII	World War II

SECTION II – TERMS

chief of mission

The principal officer (the ambassador) in charge of a diplomatic facility of the United States, including any individual assigned to be temporarily in charge of such a facility. The chief of mission is the personal representative of the President to the country of accreditation. The chief of mission is responsible for the direction, coordination, and supervision of all USG executive branch employees in that country (except those under the command of a U.S. area military commander). The security of the diplomatic post is the chief of mission's direct responsibility. Also called **COM**. (JP 1-01)

foreign assistance

Assistance to foreign nations ranging from the sale of military equipment to donations of food and medical supplies to aid survivors of natural and manmade disasters. U.S. assistance takes three forms—development assistance, humanitarian assistance, and security assistance. (JP 1-02)

intergovernmental organization

An organization created by a formal agreement (for example, a treaty) between two or more governments. It may be established on a global, regional, or functional basis for wide-ranging or narrowly defined purposes. Formed to protect and promote national interests shared by member states. Examples include the UN, NATO, and the African Union. Also called **IGO**. (JP 1-02)

international military eduction and training

Formal or informal instruction provided to foreign military students, units, and forces on a nonreimbursable (grant) basis by offices or employees of the United States, contract technicians, and contractors. Instruction may include correspondence courses; technical, educational, or informational publications; and media of all kinds. Also called **IMET**. (JP 1-02)

military assistance advisory group

A joint Service group, normally under the military command of a commander of a unified command and representing the SecDef, which primarily administers the U.S. military assistance planning and programming in the host country. Also called **MAAG**. (JP 1-02)

Military Assistance Program

That portion of the U.S. security assistance authorized by the Foreign Assistance Act of 1961, as amended, which provides defense articles and services to recipients on a nonreimbursable (grant) basis. Also called **MAP**. (JP 1-02)

other government agency

Within the context of interagency coordination, a non-DOD agency of the USG. Also called **OGA**. (JP 1-02)

status-of-forces agreement

An agreement that defines the legal position of a visiting military force deployed in the territory of a friendly state. Agreements delineating the status of visiting military forces may be bilateral or multilateral. Provisions pertaining to the status of visiting forces may be set forth in a separate agreement, or they may form a part of a more comprehensive agreement. These provisions describe how the authorities of a visiting force may control members of that force and the amenability of the force or its members to the local law or to the authority of local officials. Also called **SOFA**. (JP 1-02)

References

Army Publications

AR 360-1, *The Army Public Affairs Program*, 15 September 2000

DA Pam 27-1, *Treaties Governing Land Warfare*, 7 December 1956

DA Pam 360-512, *Code of the U.S. Fighting Force*, 1 June 1988

DA Pam 360-544, *You and the Law Overseas*, 10 November 1988

FM 3-0, *Operations*, 27 February 2008

FM 3-05.20, *(C) Special Forces Operations (U)*, 10 October 2006

FM 3-05.202, *Special Forces Foreign Internal Defense Operations*, 2 February 2007

FM 3-13, *Information Operations: Doctrine, Tactics, Techniques, and Procedures*, 28 November 2003

FM 3-61.1, *Public Affairs Tactics, Techniques and Procedures*, 1 October 2000

FM 6-22, *Army Leadership*, 12 October 2006

FM 27-10, *The Law of Land Warfare*, 18 July 1956

Department of Defense Publications

DOD Directive 5530.3, *International Agreements*, 11 June 1987

JP 1-02, *Department of Defense Dictionary of Military and Associated Terms*, 12 April 2001

JP 3-07.1, *Joint Tactics, Techniques, and Procedures for Foreign Internal Defense (FID)*, 30 April 2004

JP 3-08, *Interagency, Intergovernmental Organization, and Nongovernmental Organization Coordination During Joint Operations*, 17 March 2006

Other Publications

Bailey, Cecil E., "OPATT: The U.S. Army SF Advisers in El Salvador," *Special Warfare Magazine*, Volume 17, Issue 2, December 2004.

Builder, Carl H., *The Masks of War: American Military Styles in Strategy and Analysis*, RAND Corporation, Baltimore, MD, 1989

Foreign Assistance Act of 1961 (http://www.usaid.gov/policy/ads/faa.pdf)

International Security Assistance and Arms Export Control Act of 1976 (https://www.alt.army.mil/portal/oasaalt/documents/faa_aeca.pdf)

Ramsey, Robert D., *Advice for Advisors: Suggestions and Observations from Lawrence to the Present*, Combat Studies Institute Press, Fort Leavenworth, KS, 2006

Ramsey, Robert D., *Advising Indigenous Forces: American Advisors in Korea, Vietnam, and El Salvador*, Combat Studies Institute Press, Fort Leavenworth, KS, 2006

Special Forces Advisor's Reference Book, United States Army Special Forces Command, October 2001

This page intentionally left blank.

Index

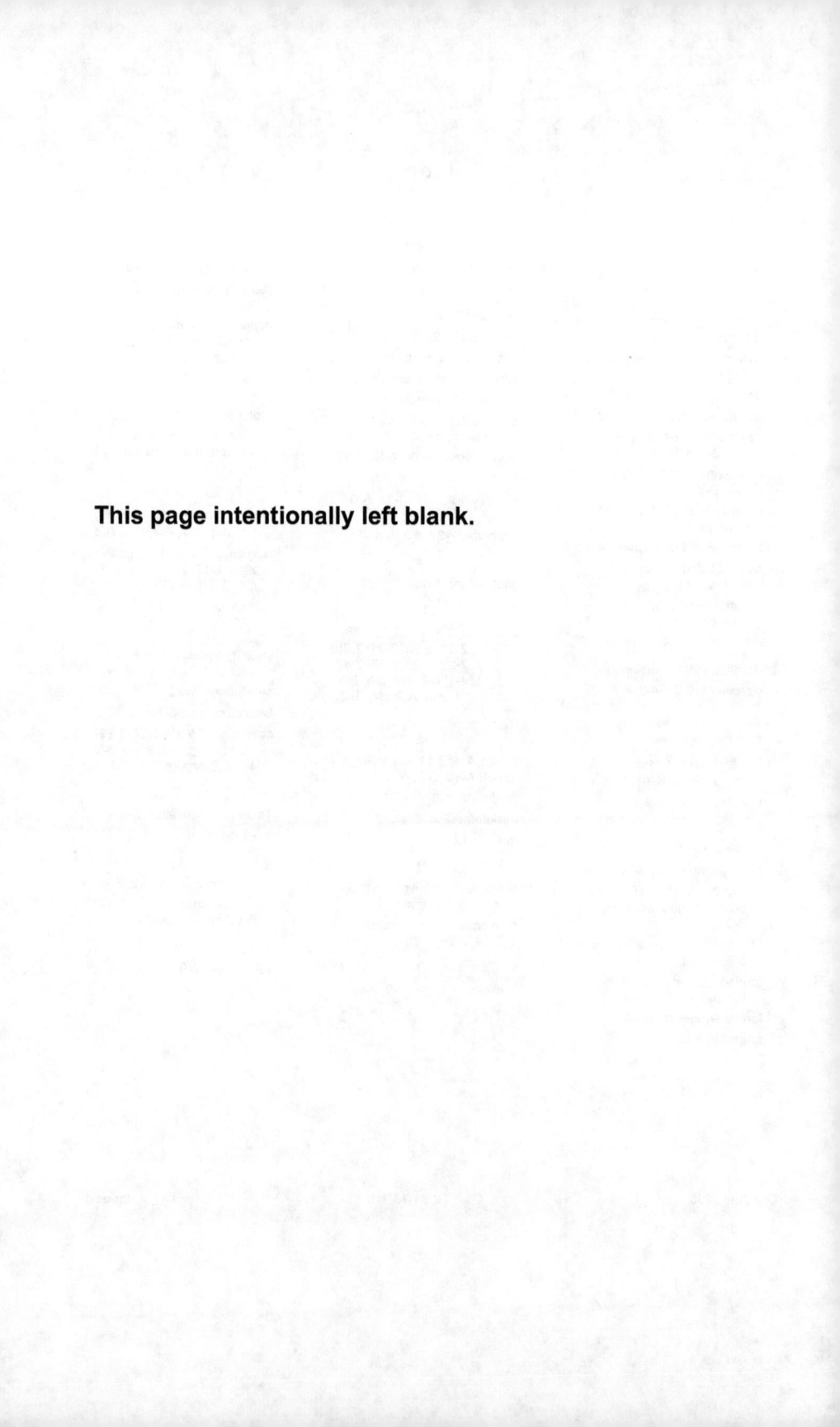

This page intentionally left blank.

TC 31-73
2 July 2008

By Order of the Secretary of the Army:

GEORGE W. CASEY JR.
General, United States Army
Chief of Staff

Official:

Joyce E. Morrow

JOYCE E. MORROW
Administrative Assistant to the
Secretary of the Army
0814001

DISTRIBUTION:

Active Army, Army National Guard, and U. S. Army Reserve: To be distributed in accordance with the initial distribution number 115988, requirements for TC 31-73.